ATLAS
of the
human
body

Vigué-Martín

ATLAS
of the
human
body

CHARTWELL
BOOKS, INC.

© 2004 Gorg Blanc
© 2004 Rebo International b.v., Lisse, The Netherlands

Director of project: Jordi Vigué
Professional consultant: Dr. Emilio Martín Orte
Illustrations: Miquel and Miriam Ferrón
Supervisor of illustrations: Marta Ribón
Cover design: Celia Valero
Pre-press service: AdAm Studio, Prague, Czech Republic
Translation of contents: Eva Šprinclová
Translation of index: Eva Šprinclová, Jan Křišťan, Kateřina Hloušková
Proofreading: Elizabeth A. Haas

This edition published in 2011 by
CHARTWELL BOOKS, INC.
A division of BOOK SALES, INC.
276 Fifth Avenue Suite 206
New York, New York 10001
USA

ISBN-10: 0 7858 2050 7
ISBN-13: 978 0 7858 2050 5

Printed in Singapore.

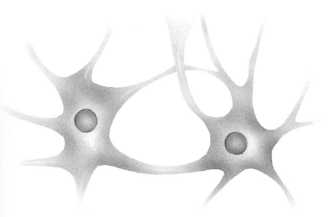

The human body is a formidable machine which carries out a complicated and diverse series of functions, including breathing, eating, reproduction, excretion of wastes and adaption to external conditions. Accomplishing each function requires the development of a specific system such as the respiratory, digestive, urinary and auditory systems. In turn, each system is composed of specific organs such as the heart, stomach, lungs and eyes.

Each organ and system depends on all the others to function correctly and contribute to the physiological balance that makes us human beings.

Like any machine, the body needs maintenance and care, both preventive and curative, to enable it to function efficiently. Although the body is very sophisticated and complex, knowledge of its composition and functions is essential to maintaining health.

This book presents a systematic overview of the human body. It is divided into chapters, each describing a specific system. The pages of each chapter are comprised of illustrations and the corresponding explanations add a wealth of detail. The objective is not only to provide an attractive book, but also one which is capable of explaining any question the reader may have. *Atlas of the Human Body* offers a wide-ranging selection of information.

The characteristics of the Atlas means it is suitable for a wide range of readers. The quality of the illustrations and their level of detail, combined with the concise and precise textual explanations, make it useful for secondary students and teachers and for many professional groups – trainers, physiotherapists, gymnasts, sportsmen, homeopaths, nurses, masseuses and many others – as well as those beginning the study of medicine or the general reader interested in health and caring for their own body.

Special care has been taken to ensure the quality of the book and to include information which may be lacking in comparable books.

We believe that *Atlas of the Human Body* represents an exciting new perspective on the field of human anatomy.

We hope that as well as providing any information that the reader might seek, the book will help people care for and understand their bodies and improve their quality of life.

Jordi Vigué

CONTENTS

DIGESTIVE SYSTEM

THE RESPIRATORY SYSTEM

THE URINARY SYSTEM

THE REPRODUCTIVE SYSTEM

THE BLOOD

GLANDULAR SYSTEM

THE NERVOUS SYSTEM

THE SENSORY SYSTEM

THE STRUCTURE OF THE HUMAN BODY

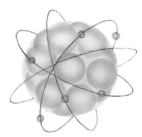

atoms

Whatever form matter adopts - solid, liquid or gas - it is always composed of tiny structural units called atoms, which are made up of a centre or nucleus containing the mass, and an orbital system containing gyrating electrons, particles with no mass and with a negative electric charge. The nucleus contains at least two other particles: neutrons, with no electric charge, and protons, with a positive electric charge. The equilibrium of the system is maintained by the balance between the positive electric charge (protons) and the negative charge (electrons). To date, 92 different electrons have been identified.

nitrogen

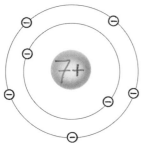

carbon

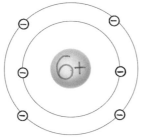

hydrogen

oxygen

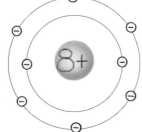

elements

The next step in the structure of matter are elements: grouping of atoms of the same type. Living matter consist of four basic elements: carbon, hydrogen, nitrogen and oxygen.

human body

The union and coordination of the different systems which make up the complex structure that is the human body.

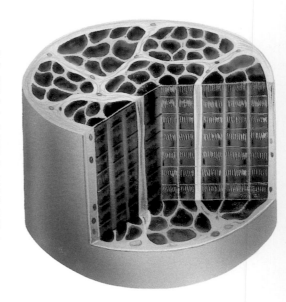

systems

The union of the different organs gives rise to the formation of different functional units which make up the macroscopic structures of the human body, each of which has an overall function: nutrition, defense, purification, support or regulation.

compounds

The union of the different elements gives rise to the compounds, whose minimum expression is known as a molecule. An example is the water molecule, formed of two atoms of hydrogen and one of oxygen. The compounds that form living matter may be organic or inorganic, depending on whether they contain carbon atoms or not. The basic organic compounds are: water, proteins, carbohydrates and fats, to which others, such as nucleic acids and steroids, are added.

organs

The different tissues combine to form organs, structures which have a specific function or functions in the human body.

cells

Different compounds (water, carbohydrates, proteins, fats, nucleic acids, etc.) combine to form cells: living organisms with complicated mechanisms of nutrition, digestion, energy production, reproduction and, in many cases, movement. Many living organisms are single-celled, but the human body, the most complex living structure, is composed of more than 100 trillion cells.

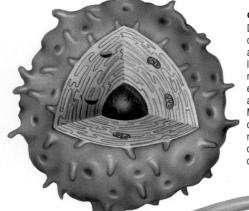

tissues

The cells of the human body combine to form more complicated anatomical elements. The main body tissues are: squamous or pavement epithelial tissue (skin and mucous); secreting epithelial tissue (exocrine and endocrine glands); connective tissue (bones, cartilages, lungs, fatty tissue, etc.); muscular tissue; blood tissue (blood); lymph tissue (lymph nodes, bone marrow, etc.); and, nervous tissue.

8

EXTERNAL ANATOMICAL FEATURES

▼ FEMALE ANTERIOR VIEW

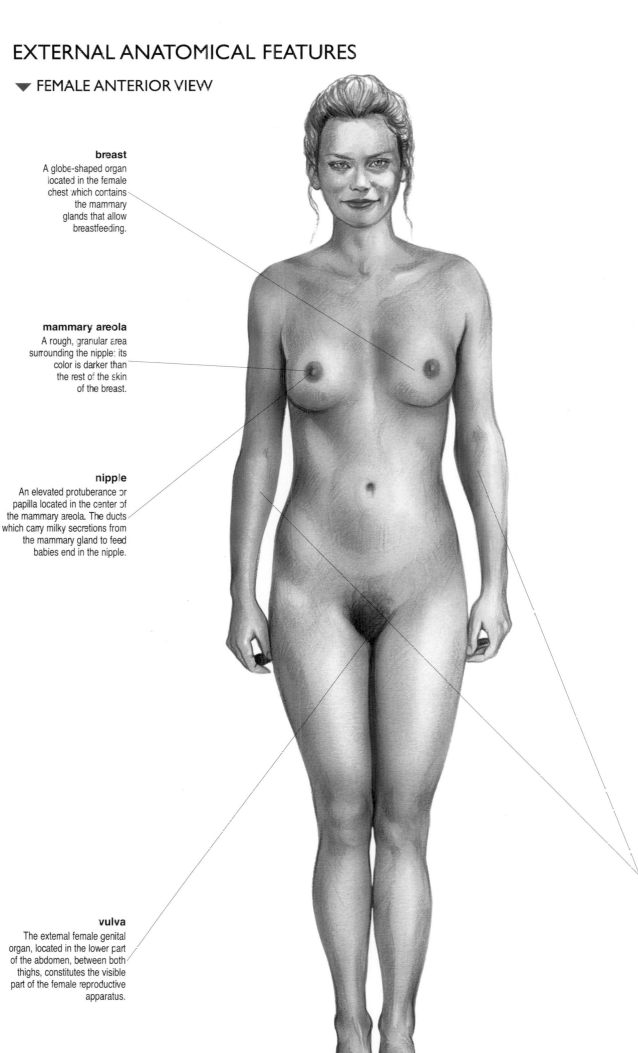

breast
A globe-shaped organ located in the female chest which contains the mammary glands that allow breastfeeding.

mammary areola
A rough, granular area surrounding the nipple; its color is darker than the rest of the skin of the breast.

nipple
An elevated protuberance or papilla located in the center of the mammary areola. The ducts which carry milky secretions from the mammary gland to feed babies end in the nipple.

vulva
The external female genital organ, located in the lower part of the abdomen, between both thighs, constitutes the visible part of the female reproductive apparatus.

head
The upper part of the human body which contains the brain and most of the sense organs.

thorax
The area of the human body located between the head and the abdomen which contains the heart and lungs and where the upper extremities are inserted.

abdomen
The area of the human body located below the thorax, which extends from the waist to the groin. It is where the lower extremities are inserted and contains most of the alimentary canal, the spleen, the urinary system and the reproductive system.

limbs
The two upper and two lower limbs, known as extremities, are inserted in the thorax and abdomen, respectively, and are composed articulated portions. They are instrumental in movement, posture and bipedalism, allowing humans to move freely and accomplish complex maneuvers, some exclusive to the human species.

9

EXTERNAL ANATOMICAL FEATURES

▼ MALE ANTERIOR VIEW

ear
The fleshy, visible part of the hearing system, located at each side of the head. The ears collect sounds and channels them towards the acoustic duct.

nose
An appendix located in the center of the face which communicates the respiratory system with the exterior.

nipple
An elevated protuberance or papilla in the center of the mammary areola.

waist
A fold or narrowing that separates the ribs from the hips.

navel
The residual scar left after tying off the fetal umbilical cord.

hip
A salient rim that is located on both sides of the abdomen below the waist and marks the hip bone.

groin
A lateral fold that ascends obliquely from the genital area and marks the area where the lower extremities are united with the abdomen.

fingernails
Accessory organs of the skin formed by hard, dense, keratinized cells that cover the posterior distal part of the fingers.

pubis
A triangular area located in the lower part of the abdomen immediately above the genitals. In adults, it is usually covered with hair.

penis
The external male genital organ which has both urinary and reproductive functions.

scrotum
A saccular structure located between the thighs and behind the penis, which contains the testicles or testes.

ankle
The joint which articulates the bones of the leg and the foot.

toes
Five small appendices located in the anterior zone of each foot.

eyes
The external organs of sight which are contained in the ocular or orbital cavities and protected by the eyelids.

mouth
An orifice located in the face that serves as the entrance to the digestive system and contains the external organs of the sense of taste.

neck
A tubular part of the body that unites the head with the thorax and through which the digestive, respiratory and nervous systems pass.

shoulder
The area where the upper limb joins the thorax. Due to its powerful musculature, the shoulder usually has a rounded aspect.

armpit
A concave area located in the inferior angle of the union of the upper limb with the thorax. In adults, the armpit is usually covered with hair.

arm
The first portion of the upper limb which extends from the shoulder and the armpit to the elbow.

flexure of the elbow
The anterior face or flexure of the joint which articulates the arm and the forearm.

forearm
The portion of the upper extremity that extends from the elbow to the wrist.

wrist
The area which joins and articulates the bones of the forearm and the hand.

hand
The distal extremity of the upper member that, thanks to the fingers, is equipped with great dexterity and contributes greatly to the distinctiveness of the human species.

fingers
The five distal extremities of each hand.

thigh
The upper part of the lower limb that extends from the groin to the knee.

knee
The middle area of the lower limbs where the thigh and leg are joined and articulated.

feet
The distal extremities of the lower limbs. They are essential for walking, posture and bipedalism.

10

EXTERNAL ANATOMICAL FEATURES

▼ MALE POSTERIOR VIEW

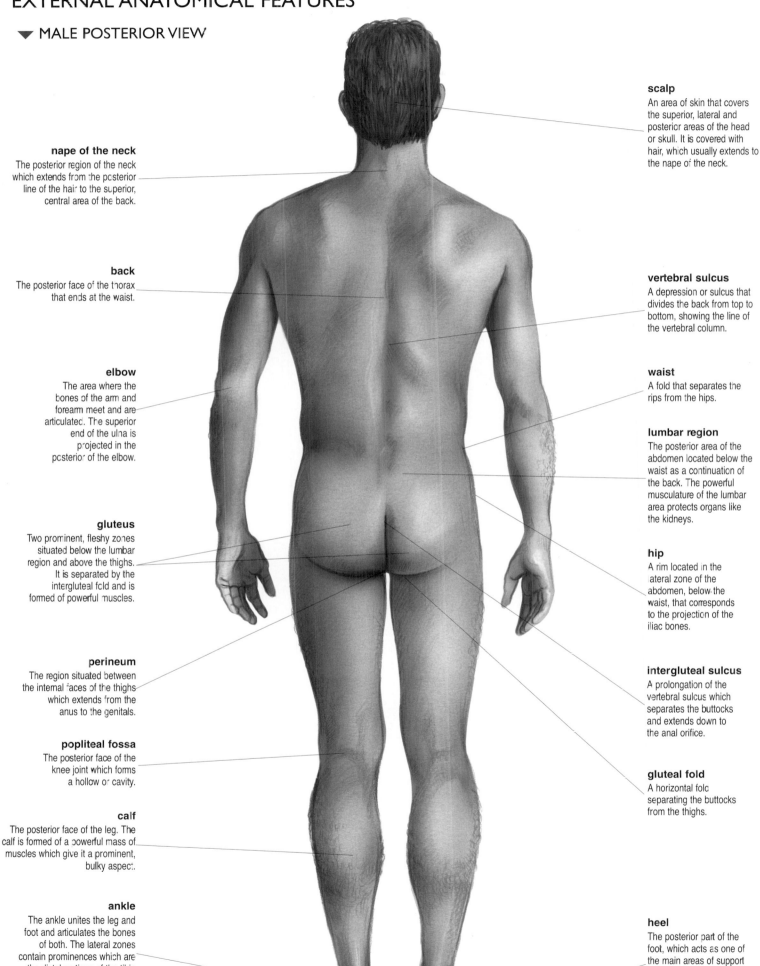

nape of the neck
The posterior region of the neck which extends from the posterior line of the hair to the superior, central area of the back.

back
The posterior face of the thorax that ends at the waist.

elbow
The area where the bones of the arm and forearm meet and are articulated. The superior end of the ulna is projected in the posterior of the elbow.

gluteus
Two prominent, fleshy zones situated below the lumbar region and above the thighs. It is separated by the intergluteal fold and is formed of powerful muscles.

perineum
The region situated between the internal faces of the thighs which extends from the anus to the genitals.

popliteal fossa
The posterior face of the knee joint which forms a hollow or cavity.

calf
The posterior face of the leg. The calf is formed of a powerful mass of muscles which give it a prominent, bulky aspect.

ankle
The ankle unites the leg and foot and articulates the bones of both. The lateral zones contain prominences which are the distal portions of the tibia and fibula and are known as the malleolus.

scalp
An area of skin that covers the superior, lateral and posterior areas of the head or skull. It is covered with hair, which usually extends to the nape of the neck.

vertebral sulcus
A depression or sulcus that divides the back from top to bottom, showing the line of the vertebral column.

waist
A fold that separates the ribs from the hips.

lumbar region
The posterior area of the abdomen located below the waist as a continuation of the back. The powerful musculature of the lumbar area protects organs like the kidneys.

hip
A rim located in the lateral zone of the abdomen, below the waist, that corresponds to the projection of the iliac bones.

intergluteal sulcus
A prolongation of the vertebral sulcus which separates the buttocks and extends down to the anal orifice.

gluteal fold
A horizontal fold separating the buttocks from the thighs.

heel
The posterior part of the foot, which acts as one of the main areas of support and balance for the body. The posterior prominence of the heel is the termination of the calcaneous bone.

11

HEAD

▼ FEMALE FRONTAL VIEW

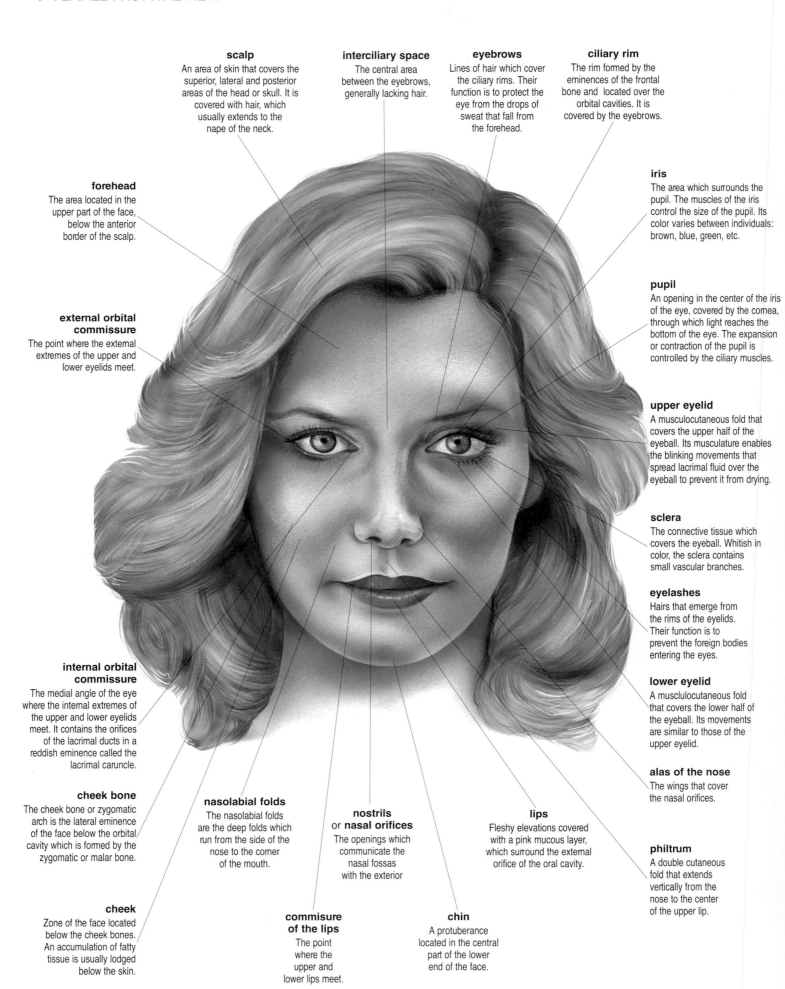

scalp
An area of skin that covers the superior, lateral and posterior areas of the head or skull. It is covered with hair, which usually extends to the nape of the neck.

interciliary space
The central area between the eyebrows, generally lacking hair.

eyebrows
Lines of hair which cover the ciliary rims. Their function is to protect the eye from the drops of sweat that fall from the forehead.

ciliary rim
The rim formed by the eminences of the frontal bone and located over the orbital cavities. It is covered by the eyebrows.

forehead
The area located in the upper part of the face, below the anterior border of the scalp.

iris
The area which surrounds the pupil. The muscles of the iris control the size of the pupil. Its color varies between individuals: brown, blue, green, etc.

external orbital commissure
The point where the external extremes of the upper and lower eyelids meet.

pupil
An opening in the center of the iris of the eye, covered by the cornea, through which light reaches the bottom of the eye. The expansion or contraction of the pupil is controlled by the ciliary muscles.

upper eyelid
A musculocutaneous fold that covers the upper half of the eyeball. Its musculature enables the blinking movements that spread lacrimal fluid over the eyeball to prevent it from drying.

sclera
The connective tissue which covers the eyeball. Whitish in color, the sclera contains small vascular branches.

eyelashes
Hairs that emerge from the rims of the eyelids. Their function is to prevent the foreign bodies entering the eyes.

internal orbital commissure
The medial angle of the eye where the internal extremes of the upper and lower eyelids meet. It contains the orifices of the lacrimal ducts in a reddish eminence called the lacrimal caruncle.

lower eyelid
A musculocutaneous fold that covers the lower half of the eyeball. Its movements are similar to those of the upper eyelid.

alas of the nose
The wings that cover the nasal orifices.

cheek bone
The cheek bone or zygomatic arch is the lateral eminence of the face below the orbital cavity which is formed by the zygomatic or malar bone.

nasolabial folds
The nasolabial folds are the deep folds which run from the side of the nose to the corner of the mouth.

nostrils or nasal orifices
The openings which communicate the nasal fossas with the exterior

lips
Fleshy elevations covered with a pink mucous layer, which surround the external orifice of the oral cavity.

philtrum
A double cutaneous fold that extends vertically from the nose to the center of the upper lip.

cheek
Zone of the face located below the cheek bones. An accumulation of fatty tissue is usually lodged below the skin.

commisure of the lips
The point where the upper and lower lips meet.

chin
A protuberance located in the central part of the lower end of the face.

12

HEAD

▼ MALE LATERAL VIEW

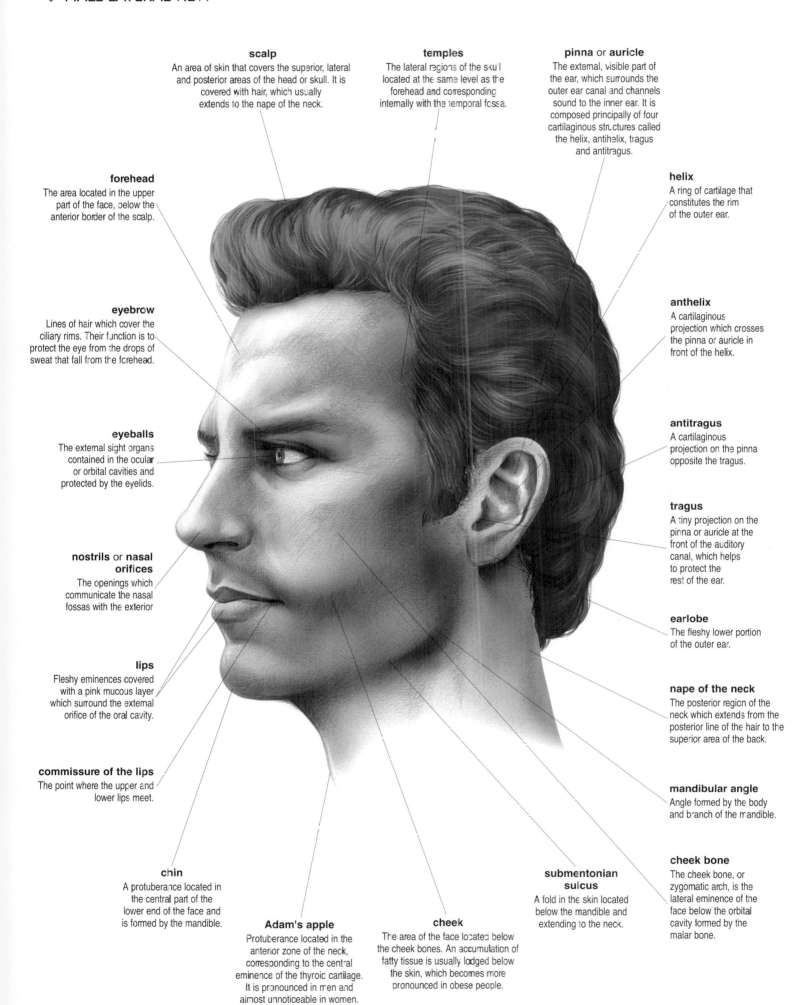

scalp
An area of skin that covers the superior, lateral and posterior areas of the head or skull. It is covered with hair, which usually extends to the nape of the neck.

temples
The lateral regions of the skull located at the same level as the forehead and corresponding internally with the temporal fossa.

pinna or auricle
The external, visible part of the ear, which surrounds the outer ear canal and channels sound to the inner ear. It is composed principally of four cartilaginous structures called the helix, antihelix, tragus and antitragus.

forehead
The area located in the upper part of the face, below the anterior border of the scalp.

helix
A ring of cartilage that constitutes the rim of the outer ear.

eyebrow
Lines of hair which cover the ciliary rims. Their function is to protect the eye from the drops of sweat that fall from the forehead.

anthelix
A cartilaginous projection which crosses the pinna or auricle in front of the helix.

eyeballs
The external sight organs contained in the ocular or orbital cavities and protected by the eyelids.

antitragus
A cartilaginous projection on the pinna opposite the tragus.

tragus
A tiny projection on the pinna or auricle at the front of the auditory canal, which helps to protect the rest of the ear.

nostrils or nasal orifices
The openings which communicate the nasal fossas with the exterior

earlobe
The fleshy lower portion of the outer ear.

lips
Fleshy eminences covered with a pink mucous layer which surround the external orifice of the oral cavity.

nape of the neck
The posterior region of the neck which extends from the posterior line of the hair to the superior area of the back.

commissure of the lips
The point where the upper and lower lips meet.

mandibular angle
Angle formed by the body and branch of the mandible.

chin
A protuberance located in the central part of the lower end of the face and is formed by the mandible.

submentonian sulcus
A fold in the skin located below the mandible and extending to the neck.

cheek bone
The cheek bone, or zygomatic arch, is the lateral eminence of the face below the orbital cavity formed by the malar bone.

Adam's apple
Protuberance located in the anterior zone of the neck, corresponding to the central eminence of the thyroic cartilage. It is pronounced in men and almost unnoticeable in women.

cheek
The area of the face located below the cheek bones. An accumulation of fatty tissue is usually lodged below the skin, which becomes more pronounced in obese people.

13

CELL STRUCTURE

nucleus
A spherical corpuscle located in the center of the cytoplasm. It contains all the genetic material of the cell, including the inherited codes which play an important role in reproduction, growth and cellular metabolism

granular endoplasmic reticulum
A complex structure formed by multiple tubular membranes that cross the cytoplasm; corpuscles called ribosomes adhere to its surface. It is considered a continuation of the nuclear membrane.

microvilli or microcilia
Prolongations emitted by the cellular membrane which enlarge the surface area of the cell, thus increasing the area of absorption, secretion, etc.

pinocytic vesicle or phagosome
A vacuole or globule formed from the cellular membrane, which, through a process known as pinocytosis, traps molecules contained in the fluid that surrounds the cell in its interior.

cellular membrane
The external layer that covers all the surface of the cell; it is elastic and permeable, allowing the products needed for the functioning of the cell to enter and waste products to be expelled. It consists of two layers of phospholipids, which are interlined with proteins and carbohydrates.

cytoplasm or protoplasm
A fluid contained in the cellular membrane, composed of water, proteins, fats and carbohydrates and different structures or organelles, each with a specific function. The liquid part of the cytoplasm is called the cytosol.

nuclear membrane
A double layer that covers the nucleus and separates it from the cytoplasm. Its porous structure allows constant communication between both.

nucleoplasm
A fluid contained in the nuclear membrane in which the internal nuclear structures float.

smooth endoplasmic reticulum
Like the granulated endoplasmic reticulum, it is formed by membranes arranged in tubular forms in the interior of the cytoplasm, but, unlike the granular reticulum, it does not have ribosomes adhering to its membrane. Its function is the synthesis of proteins, glycoproteins and lipids.

peroxisome
A corpuscle similar to the lysosome that contains enzymes, although unlike the lysosome, the peroxisome is involved in cellular metabolism through the oxidation process.

Golgi apparatus
Cavities formed of cisternae and vesicles, which are surrounded by fine membranes that unite them. Their basic function is the transport of proteins from one part of the cytoplasm to another and to the exterior of the cell.

mitochondrion
A tubular organelle composed of two membranes. The internal membrane is folder into cristae or crests. Mitochondria play an important role in cellular respiration and energy production.

ribosome
A small corpuscle that adheres to the membranes of the granular endoplasmic reticulum. In its interior, the proteins of the organism are manufactured through the combination of different amino acids.

microfilaments
The cytoplasm is furrowed by a series of microfilaments forming part of the cytoskeleton, which maintains the cell shape and aids movement.

centrioles
Two hollow, cylindrical structures, located near the nucleus, whose walls are formed by tubular systems. Their function is cellular reproduction.

lysosome
A vesicle containing many digestive enzymes that capture the nutritious substances contained in the phagosomes and digest them, manufacturing a substance that can be used by the cell and waste products that are eliminated.

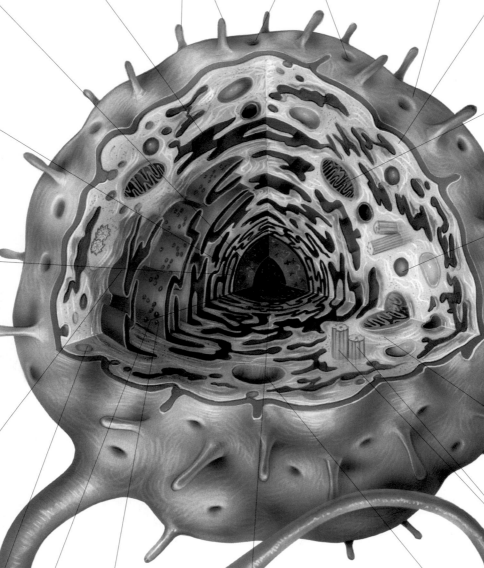

chromosomes
Thin filaments composed of long chromatin threads. They contain the genes, or genetic chains that store the specific inherited characteristics or traits of each individual (color of the eyes or the hair, glandular functions, etc.).

nucleolus
A spherical, intranuclear structure that plays an important role in cellular reproduction through the synthesis of nucleic acids.

flagellum
A propulsive structure used by many cells to enable them to move by vibration or wriggling. The spermatozoa, male sexual cells, have one long tail or flagellum that equips them with great mobility.

14

CHROMOSOMES: DNA

short arm
Shorter half of the two into which the centromere divides the chromatids.

centromere
Point at which the two identical chromatids unite to form a chromosome.

chromosomes
Structures in the cellular nucleus, visible only during the reproductive phase, when they are formed from chromatin filaments. Each chromosome is formed by two identical halves called chromatids, and thus their DNA is duplicated. Each species has a specific number of chromosomes; humans have 46, arranged in 23 pairs. Of these, 22 pairs are identical in men and women (autosomes), and there is one pair called X (female) or Y (male) chromosomes, which define the sex (gonosomes or sexual chromosomes).

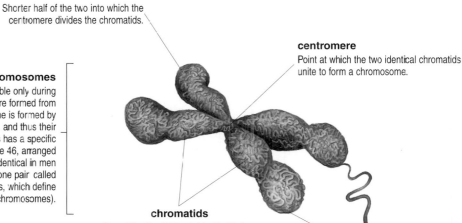

chromatids
One of the tiny (700nm), identical halves which form a chromosome. They are constituted of filaments of chromatin folded and rolled in on themselves. Each chromatid has the same DNA composition as its pair, since they will separate and contribute the same genetic load to two future cells.

long arm
Longer half of the two into which the centromere divides the chromatids.

nucleotide
One of the components of the DNA chains, nucleotides are the fragments where the phosphoric acid and deoxyribose are united with guanine, adenine, thymine or cytosine. Thus, in each DNA chain there are four different nucleotides that are repeated in a specific sequence.

nucleosomes
The double DNA helix, if unfolded, would have a length of up to 2 in. In order to make it fit in a very small space, such as the nucleus, it must fold in on itself to a remarkable degree; this is accomplished by proteins called histones that, united with the DNA chains, form the nucleosomes. The DNA chains are wrapped around the histones in two turns, shortening the length of the chain considerably.

DIAGRAM OF THE DNA CHAIN

DNA
(DEOXYRIBONUCLEIC ACID)

The chromatin filaments are formed by two structures in the form of a helix, which are constituted by a skeleton composed of phosphoric acid, and a sugar, deoxyribose to which four nitrogenated substances adhere in a specific sequence: adenine (A), guanine (G), thymine (T) and cytosine (C). In turn, the two chains are united by hydrogen bonds. The different combinations of these four elements (A, G, T and C) manufacture a code that, correctly interpreted, reveals the genetic message of the cell.

THE **23** CHROMOSOMES OF THE HUMAN KARYOTYPE

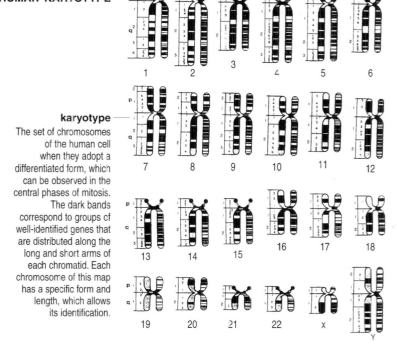

karyotype
The set of chromosomes of the human cell when they adopt a differentiated form, which can be observed in the central phases of mitosis. The dark bands correspond to groups of well-identified genes that are distributed along the long and short arms of each chromatid. Each chromosome of this map has a specific form and length, which allows its identification.

15

CELLULAR REPRODUCTION

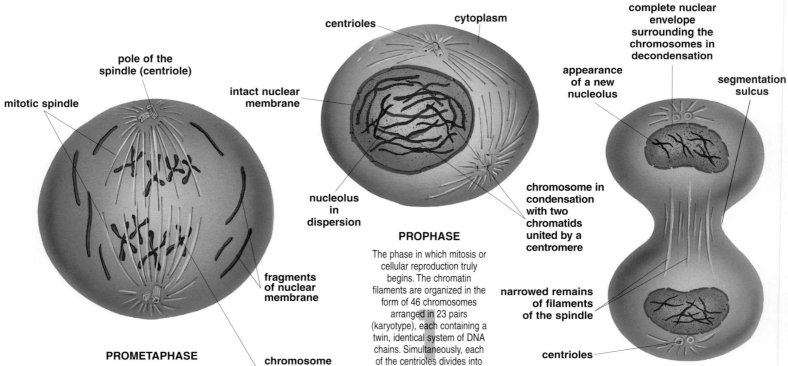

centrioles

cytoplasm

intact nuclear membrane

nucleolus in dispersion

PROPHASE

The phase in which mitosis or cellular reproduction truly begins. The chromatin filaments are organized in the form of 46 chromosomes arranged in 23 pairs (karyotype), each containing a twin, identical system of DNA chains. Simultaneously, each of the centrioles divides into two new centrioles that separate from each other and begin to migrate to opposite cellular poles.

mitotic spindle

pole of the spindle (centriole)

fragments of nuclear membrane

chromosome placed at random in active movement

PROMETAPHASE

In this stage, the centrioles have already been placed in totally opposed cellular poles, although they remain united by fibers composed of the tubules of the cytoskeleton that make up the mitotic spindle. Simultaneously, the nuclear membrane dissolves, the nucleolus disappears and the chromosomes are fixed to the fibers of the mitotic spindle by special structures of the centromere called kinetochores.

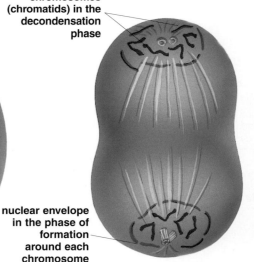

complete nuclear envelope surrounding the chromosomes in decondensation

segmentation sulcus

appearance of a new nucleolus

chromosome in condensation with two chromatids united by a centromere

narrowed remains of filaments of the spindle

centrioles

CYTOKINESIS

In the cellular cytoplasm, the mitotic spindles disappear and the cellular membrane begins to experience a progressive narrowing in its equatorial plane, the segmentation sulcus, which ends in the splitting of the mother cell into two identical cells. Inside the new nuclear membrane, the nucleoli appear. When the reproduction is finished, the cell will return to the interphase or non-reproductive period.

PHASES OF MITOSIS

Mitosis is the form of cellular reproduction which enables the cell to divide after duplicating its genetic load, so that each daughter cell receives a complete set of chromosomes.

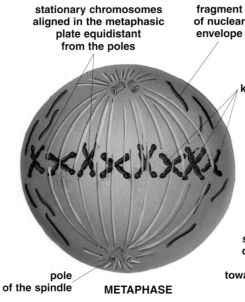

stationary chromosomes aligned in the metaphasic plate equidistant from the poles

fragment of nuclear envelope

kinetochores

pole of the spindle

METAPHASE

Once the centrioles have been placed in totally opposed planes of the cell and the chromosomes have been united to the fibers of the mitotic spindle by the kinetochores, each of the two centrioles begins to exert an attraction of equal intensity on the chromosomes and, consequently, these migrate to the equatorial plane of the cell, constituting an equatorial or metaphasic plate.

separated chromatid attracted towards the pole

ANAPHASE

During the anaphase, the attraction of the centrioles on the chromosomes intensifies, so that each pair of chromosomes is divided by its centromere into two halves and initiates a migration towards an opposed cellular pole, while the cell becomes elongated, increasing the distance between the poles.

chromosomes (chromatids) in the decondensation phase

nuclear envelope in the phase of formation around each chromosome

TELOPHASE

In this phase, the chromosomes are organized around each centriole, while a membranous structure, the future nuclear membrane, begins to form around them. At the end of the telophase, the chromosomes begin to become disorganized and adopt the diffuse chromatin form again.

BODY TISSUES

TISSUES

The cells of the human body are grouped together to form more complicated structures called tissues. These are the elements from which the different systems that form the human body are constructed. There are seven types of body tissue.

PAVEMENT EPITHELIA

simple epithelium

Tissues that cover the outside of the human body and the interior of the organic cavities, forming the skin and the mucosa, respectively. They can be formed by cells arranged in a single layer (simple epithelium), in several layers (stratified epithelium) or in stepped form (pseudostratified epithelium).

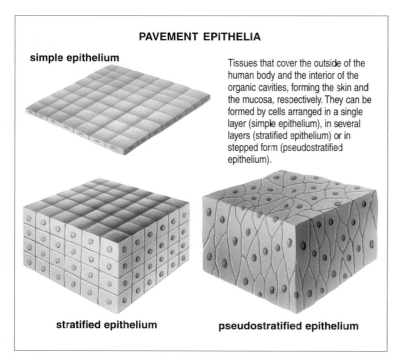

stratified epithelium

pseudostratified epithelium

GLANDULAR EPITHELIA

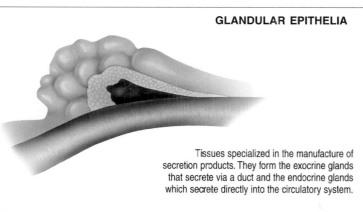

Tissues specialized in the manufacture of secretion products. They form the exocrine glands that secrete via a duct and the endocrine glands which secrete directly into the circulatory system.

CONNECTIVE TISSUE

There are various types of connective tissue, also known as conjunctive tissue.

LOOSE TISSUE

The matrix on which most organs are constructed (liver, alimentary canal, lungs, etc.). It forms part of the internal membranes and fills the space between them; it is formed by cells called fibroblasts, a fundamental substance composed of water, mineral sugars and salts, and fibers of collagen, reticulin and elastin.

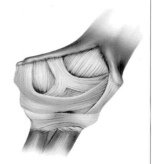

DENSE TISSUE

Dense tissue acts as a support and forms the structure of bones, tendons, ligaments and blood vessels. Its structure is similar to that of loose connective tissue, but the proportion of constituent fibers varies.

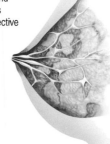

ADIPOSE TISSUE

Adipose tissue is the body's fat store and is an important reserve of energy and a protective cushion for the internal organs. It is formed by cells rich in fatty material, the adipocytes.

BLOOD

This tissue is the body's method of transporting the substances necessary for the sustenance of its cells to all parts of the body and of ridding the organs of metabolic waste products. It is composed of a liquid (plasma) and solids (blood cells).

LYMPHOID TISSUE

The tissue specializing in the production of cells which form the body's defense mechanisms (lymphocytes, plasmacytes, etc.), fighting foreign bodies such as bacteria and viruses. It is found in the lymphoid organs, which are the lymph nodes, the bone marrow and the tonsils.

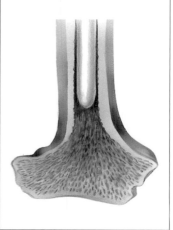

NERVOUS TISSUE

A specialized tissue that permits the transmission of the neuroelectrical impulses which constitute the basis of all the governing functions of the nervous system, allowing the brain to transmit its orders. The brain, cerebellum, spinal marrow and all the nerves of the organism are composed of nervous tissue.

MUSCULAR TISSUE

The tissue that forms muscles which, through their contraction, are able to develop mechanical functions. Smooth muscular tissue is contracted involuntarily and is found in internal organs such as the intestine, uterus and arteries. Striated muscular tissue contracts voluntarily and is found in the muscles of the extremities, neck, thorax, abdomen, etc.

17

MICROSCOPIC STRUCTURE OF THE SKIN

epidermis
The external of the three layers of the skin. It consists of five strata in which the epidermic cells, the keratinocytes, gradually evolve and harden progressively by the process known as keratinization.

dermis
The layer of the skin located below the epidermis and composed of loose connective tissue and fibrous tissue. It contains many nerve terminations and blood vessels. In this layer, the sudoriferous glands, the sebaceous glands and the roots of the hair and different types of cells such as fibroblasts, histocytes and mastocytes are located.

hypodermis
The deepest layer of the skin, located below the dermis. It is formed of loose connective tissue and contains abundant adipose tissue, which acts as a cushion for the organs below (muscles, bones, viscera, etc.), from which it is separated by the subcutaneous cellular tissue, the deepest portion of the hypodermis.

basal layer
Also known as the germinative layer, it is located in the deepest part and continuously produces new keratinocytes.

spinous layer
Located above the basal layer and composed of continuously multiplying keratinocytes.

granular layer
Formed by epidermic cells that initiate their cornification or hardening.

clear layer
The clear layer only exists in zones of very compact skin and is formed by flattened, dead keratinocytes.

horny layer
Superficial layer of the epidermis, where the keratinized eperdermic cells are shed and replaced by others. The soles of the feet and the palms of the hands have a thicker layer.

pores
Small openings that sometimes coincide with the superior end of the excretory channel of a sudoriferous gland or with the birth of a hair.

keratinocytes
The cells that form the epidermis. They originate in the basal layer and evolve continuously to terminate as dead cells shed by the corneous layer.

dermal papillae
Superior part of the dermis, formed by small, nipple-like protusions which extend into the epidermis.

blood capillaries

epidermis

dermis

hypodermis

Pacini's corpuscle
Nervous terminations located in the deepest part of the dermis that detect the deepest tactile sensations.

Meissner's corpuscle
Nervous terminations of the dermis that detect superficial tactile sensations. They are very abundant in the fingertips.

Ruffini's corpuscle
Nervous terminations of the dermis which specialize in detecting heat.

Krause's corpuscle
Nervous terminations of the dermis that detect cold sensations.

sudoriferous gland
Glandular structures in the form of a twisted tubule, which specialize in the secretion of sweat. They are located in the sinus of the dermis and expel their secretions through a duct that opens to the epidermis through the pores.

erector muscle of the hair
A thin muscle that unites the base of the hair follicle with the epidermis. Its function is to facilitate the erection of the hair in situations of cold, stress, etc.

Langerhan's cells
Cells located in the spinous layer between the keratinocytes. Morphologically, they resemble melanocytes.

melanocytes
Cells located between the keratinocytes of the basal layer. Their function is to synthesize melanin, the substance responsible for skin and hair color, and to filter solar rays.

hair follicle
A saccular structure which contains the hair and the sebaceous glands.

ACCESSORY ORGANS OF THE SKIN: THE HAIR FOLLICLE

CROSS-SECTION OF A FOLLICLE

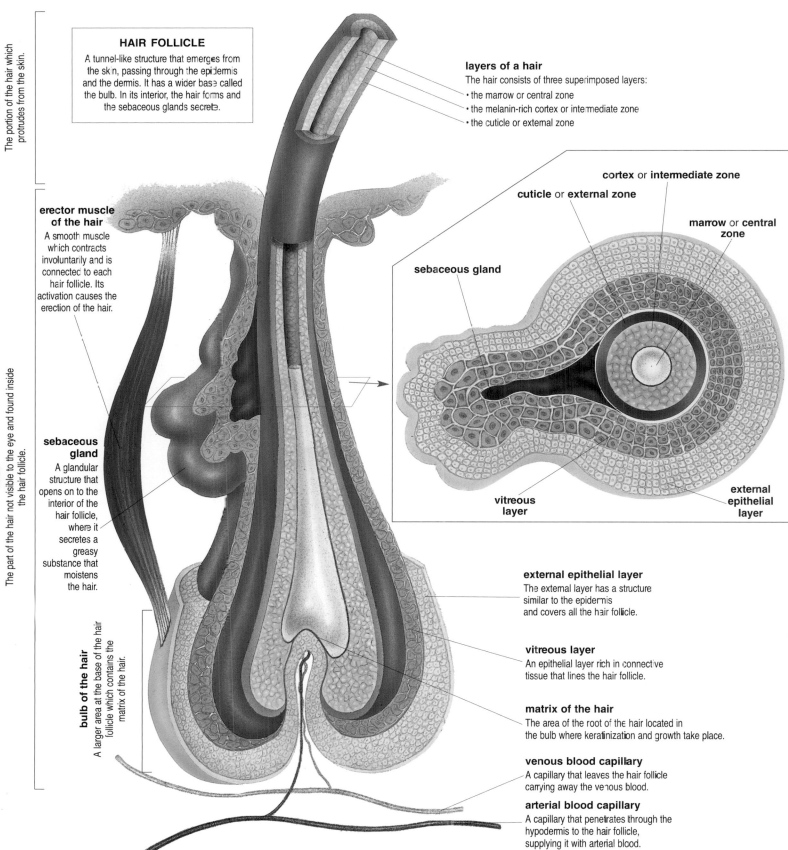

stem
The portion of the hair which protrudes from the skin.

HAIR FOLLICLE
A tunnel-like structure that emerges from the skin, passing through the epidermis and the dermis. It has a wider base called the bulb. In its interior, the hair forms and the sebaceous glands secrete.

layers of a hair
The hair consists of three superimposed layers:
• the marrow or central zone
• the melanin-rich cortex or intermediate zone
• the cuticle or external zone

erector muscle of the hair
A smooth muscle which contracts involuntarily and is connected to each hair follicle. Its activation causes the erection of the hair.

root
The part of the hair not visible to the eye and found inside the hair follicle.

sebaceous gland
A glandular structure that opens on to the interior of the hair follicle, where it secretes a greasy substance that moistens the hair.

bulb of the hair
A larger area at the base of the hair follicle which contains the matrix of the hair.

cuticle or **external zone**

cortex or **intermediate zone**

marrow or **central zone**

sebaceous gland

vitreous layer

external epithelial layer

external epithelial layer
The external layer has a structure similar to the epidermis and covers all the hair follicle.

vitreous layer
An epithelial layer rich in connective tissue that lines the hair follicle.

matrix of the hair
The area of the root of the hair located in the bulb where keratinization and growth take place.

venous blood capillary
A capillary that leaves the hair follicle carrying away the venous blood.

arterial blood capillary
A capillary that penetrates through the hypodermis to the hair follicle, supplying it with arterial blood.

19

ACCESSORY ORGANS OF THE SKIN: FINGERNAIL

body of the nail
or ungueal limb
The visible part of the fingernail located on the posterior face of the distal extremity of the fingers. It is hard and horny and is formed of keratinized epithelial cells.

paraungueal fold
A fold of skin which surrounds the lateral parts of the body of the nail. The fold can allow infections, known as panaris infections, to enter.

lunula
An area of the nail with a clearer color and semicircular edge located at the base of the body of the nail.

cuticle
Whitish, soft, membranous lamina that surrounds the body of the nail body at its base, in the area of the lunula, and separates it from the surrounding skin.

ungueal root
The newest part of the nail which is contained in the matrix.

matrix of the nail
An area located under the skin of the ends of the fingers which contains the epithelial cells which, through the process of keratinization, form the nail.

nail bed
The portion of skin on which the nail rests and which serves as a base.

free edge
The distal end of the body of the nail that, due to its growth, is continually approaching the end of the finger.

perionychium
An area of skin that separates the nail bed of the epidermis from the fingertip.

striae
of the nails
Fine, whitish lines which sometimes appear on the body of the nail due to stratification defects in the keratinized cells that form the nail.

distal phalange
The last of the phalangeal bones which supports the structure of the nail and finger.

fingertip
The area located at the anterior distal ends of the fingers. It contains characteristic cutaneous sulci called dermatoglyphs which are responsible for fingerprints.

subcutaneous fat
An accumulation of adipose tissue located under the layers of the skin, specifically in contact with the hypodermis, for which it forms the cushioning.

20

ACCESSORY ORGANS OF THE SKIN: SUDORIFEROUS GLANDS

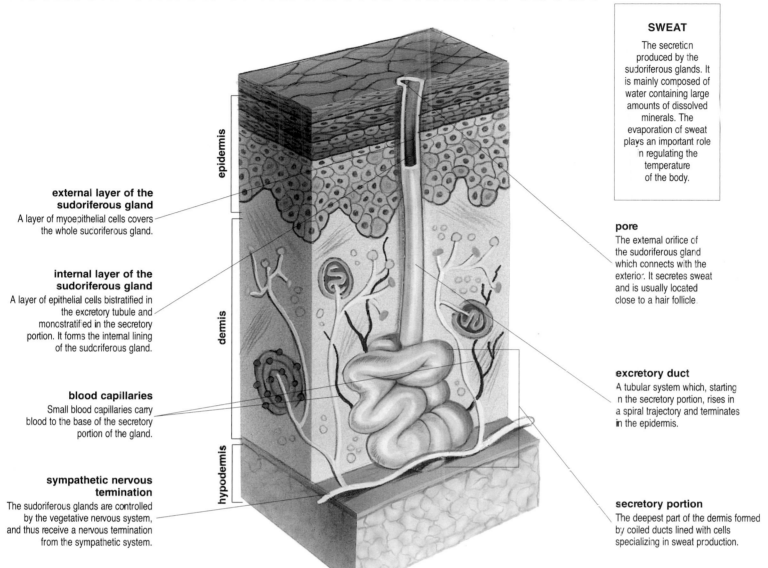

SWEAT

The secretion produced by the sudoriferous glands. It is mainly composed of water containing large amounts of dissolved minerals. The evaporation of sweat plays an important role in regulating the temperature of the body.

external layer of the sudoriferous gland

A layer of myoepithelial cells covers the whole sudoriferous gland.

internal layer of the sudoriferous gland

A layer of epithelial cells bistratified in the excretory tubule and monostratified in the secretory portion. It forms the internal lining of the sudoriferous gland.

blood capillaries

Small blood capillaries carry blood to the base of the secretory portion of the gland.

sympathetic nervous termination

The sudoriferous glands are controlled by the vegetative nervous system, and thus receive a nervous termination from the sympathetic system.

epidermis

dermis

hypodermis

pore

The external orifice of the sudoriferous gland which connects with the exterior. It secretes sweat and is usually located close to a hair follicle.

excretory duct

A tubular system which, starting in the secretory portion, rises in a spiral trajectory and terminates in the epidermis.

secretory portion

The deepest part of the dermis formed by coiled ducts lined with cells specializing in sweat production.

21

COMPOSITION OF SWEAT

- water: 98%
- total nitrogen: 25-60 gr/100 cc
- urea: 10-575 mg/100 cc
- chlorine: until 40 mlEq/l
- sodium: 10-60 mlEq/l
- potassium: 3-10 mlEq/l
- lactic acid: 45-450 mg/100 cc

- uric acid: 0,7-2,5 mg/100 cc
- pyruvic acid: 4,4 mg/100 cc
- tyrosine: 3,2 mg/100 cc
- threonine: 5,5 mg/100 cc
- arginine: 13,5 mg/100 cc
- histidine: 8 mg/100 cc

DISTRIBUTION OF THE SUDORIFEROUS GLANDS

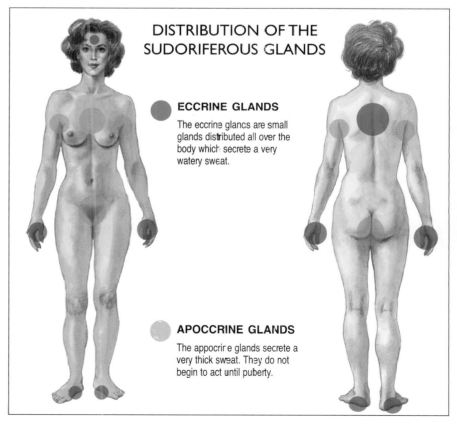

ECCRINE GLANDS

The eccrine glands are small glands distributed all over the body which secrete a very watery sweat.

APOCCRINE GLANDS

The appocrine glands secrete a very thick sweat. They do not begin to act until puberty.

MUSCULAR SYSTEM

▼ GENERAL ANTERIOR VIEW

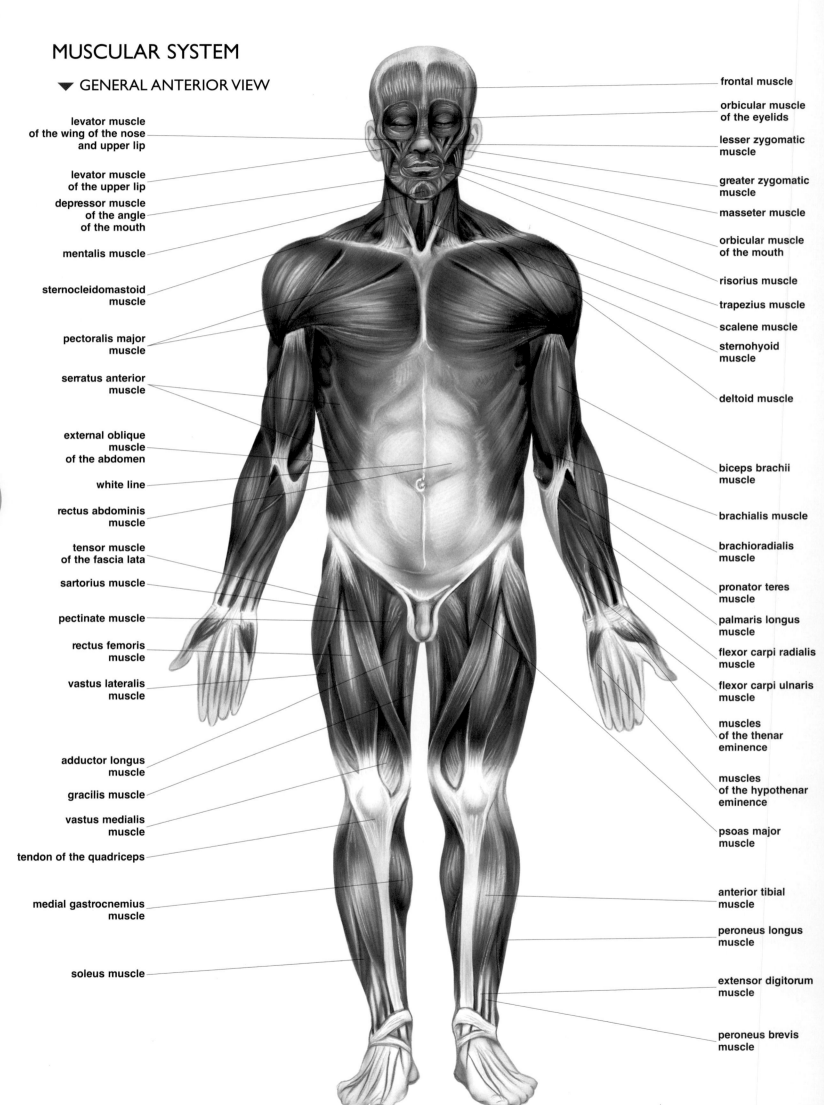

levator muscle
of the wing of the nose
and upper lip

levator muscle
of the upper lip

depressor muscle
of the angle
of the mouth

mentalis muscle

sternocleidomastoid
muscle

pectoralis major
muscle

serratus anterior
muscle

external oblique
muscle
of the abdomen

white line

rectus abdominis
muscle

tensor muscle
of the fascia lata

sartorius muscle

pectinate muscle

rectus femoris
muscle

vastus lateralis
muscle

adductor longus
muscle

gracilis muscle

vastus medialis
muscle

tendon of the quadriceps

medial gastrocnemius
muscle

soleus muscle

frontal muscle

orbicular muscle
of the eyelids

lesser zygomatic
muscle

greater zygomatic
muscle

masseter muscle

orbicular muscle
of the mouth

risorius muscle

trapezius muscle

scalene muscle

sternohyoid
muscle

deltoid muscle

biceps brachii
muscle

brachialis muscle

brachioradialis
muscle

pronator teres
muscle

palmaris longus
muscle

flexor carpi radialis
muscle

flexor carpi ulnaris
muscle

muscles
of the thenar
eminence

muscles
of the hypothenar
eminence

psoas major
muscle

anterior tibial
muscle

peroneus longus
muscle

extensor digitorum
muscle

peroneus brevis
muscle

22

MUSCULAR SYSTEM

▼ POSTERIOR GENERAL VIEW

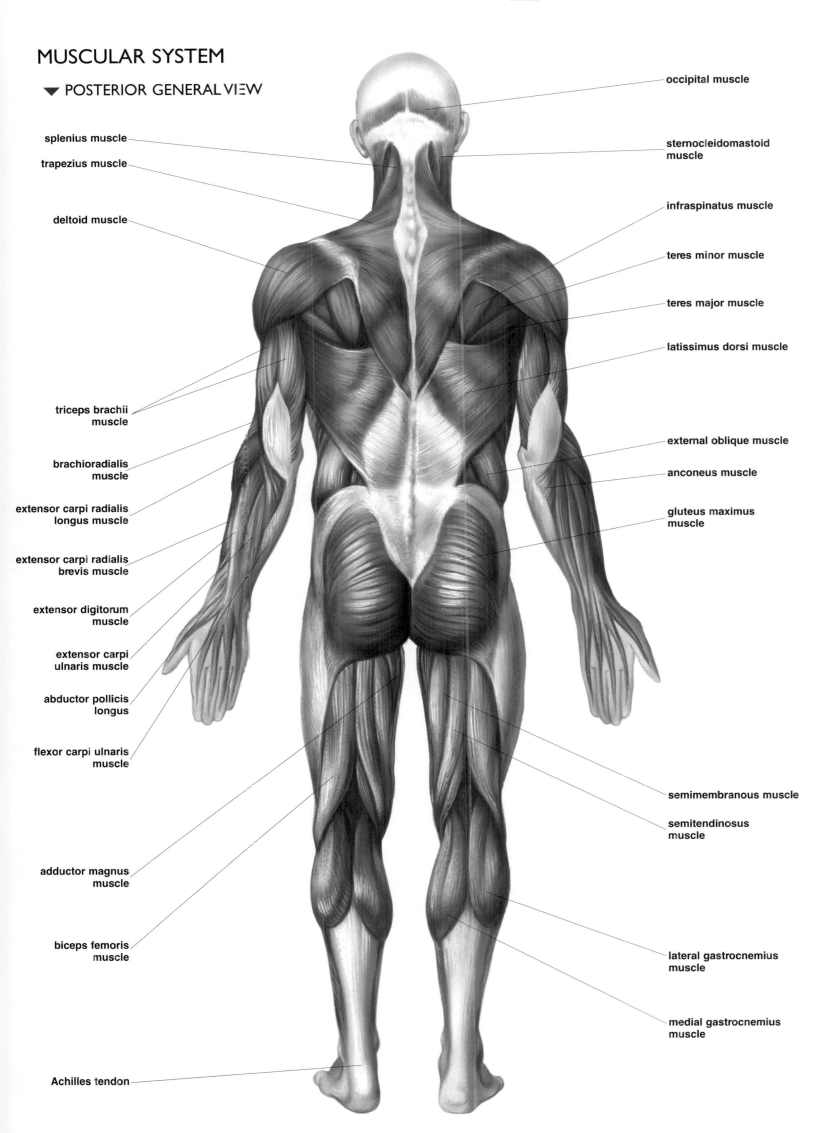

splenius muscle

trapezius muscle

deltoid muscle

triceps brachii
muscle

brachioradialis
muscle

extensor carpi radialis
longus muscle

extensor carpi radialis
brevis muscle

extensor digitorum
muscle

extensor carpi
ulnaris muscle

abductor pollicis
longus

flexor carpi ulnaris
muscle

adductor magnus
muscle

biceps femoris
muscle

Achilles tendon

occipital muscle

sternocleidomastoid
muscle

infraspinatus muscle

teres minor muscle

teres major muscle

latissimus dorsi muscle

external oblique muscle

anconeus muscle

gluteus maximus
muscle

semimembranous muscle

semitendinosus
muscle

lateral gastrocnemius
muscle

medial gastrocnemius
muscle

23

EXTERNAL AND INTERNAL STRUCTURE OF A STRIATED MUSCLE

STRIATED MUSCLE

The striated or skeletal muscles are those that move voluntarily, following a conscious order from the brain. They are transmitted through the somatic nerves and are attached to the different parts of the skeleton, allowing movement.

insertion points
The points at which the tendons are fixed to the skeleton, allowing muscular contractions to be transmitted to bones, cartilages or joints to enable movement.

tendons
Found at the extremes of almost all striated muscles, tendons serve as fixation elements between the muscles and skeleton. They are formed of pearly-colored, fibrous connective tissue.

containment aponeurosis
Membranous sheaths which surround striated muscles and separate the different muscular groups formed of fibrous connective tissue similar to that of the tendons.

muscular belly
The most voluminous part of the muscle, almost always located in the center of the muscle.

blood vessels
The muscles are served by arteries that carry oxygenated blood to them and penetrate them through fine arterial capillaries and venous capillaries carrying deoxygenated blood away to the venous network.

somatic nerve
The brain sends nervous impulses which stimulate voluntary movements to the muscle through the somatic nerve.

myofibrils
Small cylindrical filaments measuring 1-2 micrometers in diameter. Each muscular fiber contains thousands.

endomysium
Very fine layer of reticular fibers which originate in the perimysium and surround each of the muscular fibers composing the muscle.

epimysium
Membranous connective tissue that surrounds the muscle and whose prolongations form part of the tendons.

sarcolemma
The plasma membrane that surrounds each cell or muscular fiber.

blood capillaries
Small capillaries which provide the blood flow to the muscular fibers through the perimysium.

muscular fibers
Cells or structural units arranged longitudinally within the muscle. Their diameter ranges between 18-80 microns. They contain a fluid medium called sarcoplasm and myofibrils.

perimysium
Membranous walls that originate in the epimysium and surround a fascicle or packet of muscular fibers.

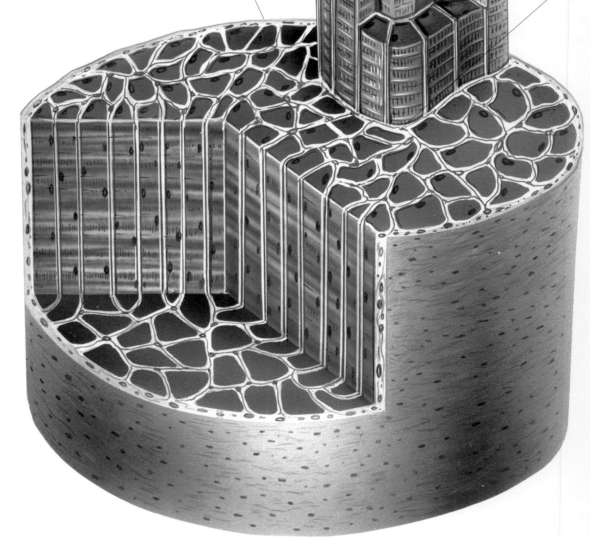

EXTERNAL AND INTERNAL STRUCTURE OF A SMOOTH MUSCLE

SMOOTH MUSCLES

Also called visceral muscles, smooth muscles move involuntarily, following automatic impulses generated in the central nervous system and transmitted through the vegetative or autonomous nervous system, without conscious thought. The smooth muscles are found in the walls of the visceral organs such as the blood vessels, intestine, bronchi, etc. and in the skin and eyes. They allow these organs to function equally during sleep or waking.

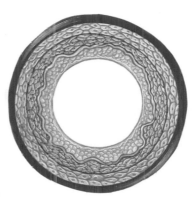

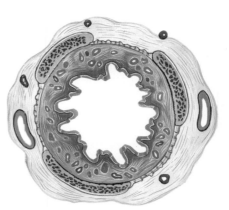

prostate and vesicle musculature
The walls of the bladder and prostate are formed from a muscular layer whose contraction allows the mechanisms triggering ejaculation or urination to occur. The bladder is also equipped with striated muscles that allow the flow of the urine to be controlled voluntarily.

musculature of the arterial walls
The musculature of the walls of the arteries allows them to change diameter in response to the blood flow and changes in the blood pressure.

musculature of the bronchial walls
The walls of the bronchi are equipped with a muscular layer that, through relaxation or contraction, allows the bronchi to widen or narrow, allowing greater or smaller air intake to the pulmonary alveoli.

25

smooth muscular fiber
The smooth muscle is formed by bundles of fusiform cells of 80-200 um in length, generally arranged in layers, especially in the walls of hollow organs (intestine, blood vessels, bronchi, etc.). It is also found in the connective tissue that lines organs such as the prostate, or forming individualized units such as the erector muscles of the hair or the muscles of the iris.

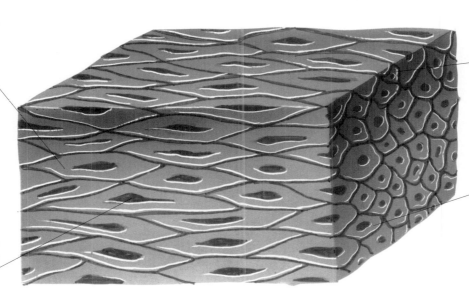

plasma membrane
Fine membrane that surrounds smooth muscle fibers. It contains a network of reticular fibers which join the muscle fibers.

sarcoplasm
Cytoplasm of smooth muscular fiber cells, which contain many tiny myofibrils, arranged irregularly and visible only under the electron microscope. The myofibrils are composed of actin and myosin and are responsible for muscular contraction.

nucleus
The cells of the smooth musculature have a single nucleus normally located in the center of the cytoplasm.

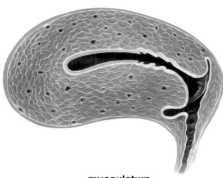

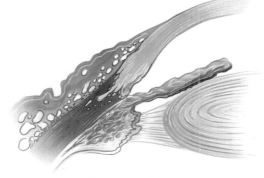

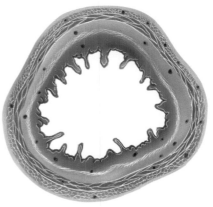

musculature of the uterine wall
The mechanisms of childbirth are possible due to the contractions of the powerful musculature of the uterine wall, triggered by hormonal stimuli.

ciliary musculature of the eye
Surrounding the lens of the eye, the ciliary muscles contract or relax, allowing the shape of the lens to change and thus optimizing the vision.

musculature of the intestinal walls
The contraction of this musculature causes peristaltic movements that allow the nutritional bolus to advance through the different segments of the alimentary canal.

SKULL AND FACE: SUPERFICIAL MUSCLES

▼ FRONTAL VIEW

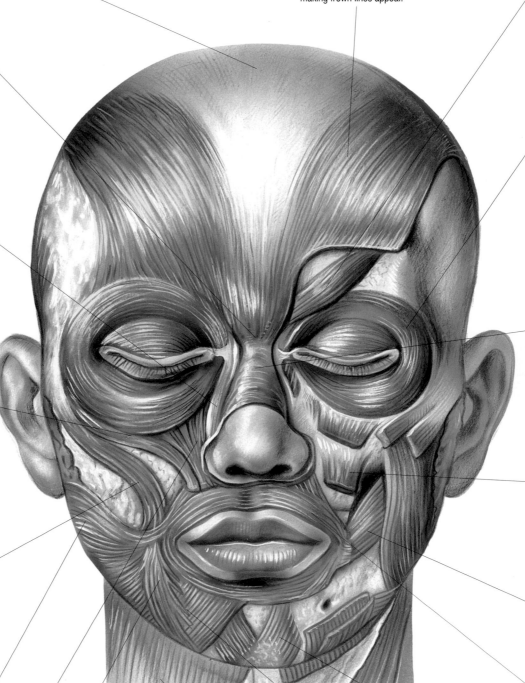

galea aponeurotica
A fibrous membrane that surrounds the superior area of the skull and is firmly joined to the skin that covers it, allowing it to slide over the bone. It serves as the origin of the various cutaneous muscles of the skull.

frontal muscle
A muscle that extends under the skin of the forehead from the galea aponeurotica to the superior border of the orbit. When contracted, it tenses the galea aponeurotica, but also intervenes in facial gestures such as raising the eyebrows and making frown lines appear.

superciliary muscle
A small, thin facial muscle located below the orbicular muscle of the eyelids and the frontal, in the internal area of the superciliary arch. Its contraction allows frowning.

pyramidal muscle
A facial muscle which extends vertically along the dorsal area of the nose, from the internal ciliary area to the cartilages and bones that form the nasal skeleton. Its contraction causes the appearance of cutaneous folds between the eyebrows.

levator muscle of the ala of the nose and upper lip
A facial muscle inserted in the internal area of the superior maxillary bone. From there, it divides into two fascicles: one goes to the skin of the alas of the nose and the other goes to the skin of the upper lip. Its contraction elevates the ala of the nose, expands the nasal orifice and causes the upper lip to protrude upwards.

orbicular muscle of the eyelids
A circular facial muscle surrounding the palpebral opening. It extends from the internal angle of the eye to the external angle, attached to the skin of the eyelids. It permits opening and closing the eyelid, allowing gestures such as winking and blinking.

nasal muscle
Also known as the transverse muscle of the nose. A facial muscle that extends from the median line of the nasal cartilages to the skin that covers the alas of the nose. When contracted, it causes the nasal orifices to narrow and vertical facial folds to appear.

lesser zygomatic muscle
A facial muscle inserted in the area of the cheek, the zygomatic bone and the skin of the upper lip, which it elevates and turns outwards when contracted.

levator muscle of the angle of the mouth
Also known as the *canine muscle*, it extends from the zygomatic bone to the skin of the commissure of the lips, which it elevates when contracted.

greater zygomatic muscle
A long, thin facial muscle that goes from the cheek next to the zygomatic bone to the skin of the commissure of the lips. When contracted, it elevates the commissure, complementing the action of the levator muscle at the angle of the mouth.

buccinator muscle
A facial muscle inserted in the skin surrounding the commissure of the lips which extends across the internal face of the cheeks to the superior border of the mandible and the inferior border of the maxilla. It widens the commissure of the lips transversally, but it also collaborates with other muscles in blowing, whistling or chewing.

risorius muscle
A facial muscle which, elevating the commissure of the lips upwards and outwards, allows smiling. It is inserted in the internal part of the skin of the parotid area and, from there, its fibers converge towards the commissure of the lips.

levator muscle of the upper lip
A facial muscle that stretches from the area of the maxilla located under the orbit to the upper lip, elevating the central part of the upper lip when contracted.

depressor muscle of the angle of the mouth
Also called the *triangular muscle of the lips*, due to its shape. It is attached to the inferior border of the mandible and the superior vertex located in the skin of the commissure of the lips. Its contraction allows the commissure of the lips to move downwards, permitting expressions of disgust or sadness.

platysma
A facial muscle that extends over the lateral part of the neck. Located in a very superficial position, right under the skin, it extends from near the lower lip and the chin to the skin that covers the clavicle. When contracted, it forces the skin of the chin and the lower lip downwards, collaborating with the depressor muscle at the angle of the mouth to form expressions of loathing or sadness.

depressor muscle of the lower lip
A muscle which is inserted in the inferior border of the bone of the chin and its fibers are fixed in the skin that covers the lower lip. This muscle allows the lower lip to turn downwards and outwards.

mentalis muscle
Small facial muscle located in the lateral part of the chin. It is inserted in the external face of the bone that forms the chin and terminates in the skin of this area so that its contraction elevates the chin.

orbicular muscle of the mouth
An elliptical facial muscle extending from one commissure of the lips to the other by means of two fascicles, one superior and the other inferior, which cross both lips internally, leaving the buccal opening in the middle. In the commissures, the muscle is inserted in the skin of the area and in the corresponding maxillary bones. It allows the opening and closing of the mouth and collaborates with other muscles to produce blowing, sucking and whistling actions, among others.

26

SKULL AND FACE: SUPERFICIAL MUSCLES

▼ LATERAL VIEW

temporal muscle
A wide, fan-shaped muscle extending from the temporal fossa to the coronoid process of the mandible which acts to raise the mandible and close the jaws, permitting the action of chewing.

superciliary muscle
A small, thin facial muscle located below the orbicular muscle of the eyelids and the frontal, in the internal area of the superciliary arch. Its contraction allows frowning.

orbicular muscle of the eyelids
A circular facial muscle which surrounds the palpebral opening. It extends from the internal angle of the eye to the external angle, attached to the skin of the eyelids. It permits the opening and closing of the eyelid, allowing gestures such as winking and blinking.

levator muscle of the ala of the nose and upper lip
A facial muscle inserted in the internal area of the superior maxillary bone. From there it divides into two fascicles: one goes to the skin of the alas of the nose and the other goes to the skin of the upper lip. Its contraction elevates the ala of the nose, expands the nasal orifice and causes the upper lip to protrude upwards.

orbicular muscle of the mouth
An elliptical facial muscle that extends from one commissure of the lips to the other by means of two fascicles, one superior and the other inferior, that cross both lips internally, leaving the buccal opening in the middle. In the commissures, the muscle is inserted in the skin of the area and in the corresponding maxillary bones. It allows the opening and closing of the mouth and it collaborates with other muscles to produce blowing, sucking and whistling actions, among others.

lesser zygomatic muscle
A facial muscle which is inserted in the area of the cheek of the zygomatic bone on one side and in the skin of the upper lip, which it elevates and turns outwards when contracted, on the other

frontal muscle
A muscle that extends under the skin of the forehead, from the galea aponeurotica to the superior border of the orbit. When contracted, it tenses the galea aponeurotica, but also intervenes in facial gestures such as raising the eyebrows and making frown lines appear.

depressor muscle of the lower lip
A muscle which is inserted in the inferior border of the bone of the chin and whose fibers are fixed in the skin covering the lower lip. This muscle allows the lower lip to turn downwards and outwards.

buccinator muscle
A facial muscle inserted in the skin surrounding the commissure of the lips and extending across the internal face of the cheeks to the superior border of the mandible and the inferior border of the maxilla. Its main action is to widen the commissure of the lips transversally, but it also collaborates with other muscles in blowing, whistling or chewing.

temporal fascia
A fibrous lamina that covers the temporal fossa and surrounds the temporal muscle.

risorius muscle
A facial muscle which, by elevating the commissure of the lips upwards and outwards, allows smiling. It is inserted in the internal part of the skin of the parotid area, from where its fibers converge towards the commissure of the lips.

galea aponeurotica
A fibrous membrane that surrounds the superior area of the skull and is firmly joined to the skin that covers it, allowing it to slide over the bone. It serves as the origin of the various cutaneous muscles of the skull.

masseter muscle
One of muscles used in chewing, the masseter muscle consists of two fascicles that go from the zygomatic arch of the face bones to the angle and the ascending branch of the mandible. It serves to elevate the mandible and is thus essential for chewing.

superior auricular muscle
A flat, almost atrophic muscle, located above the auricular pavilion. It goes from the lateral border of the galea aponeurotica to the superior area of the auricular cartilages, which it moves slightly upwards when contracted. There are also anterior and posterior auricular muscles.

occipital muscle
A flat muscle composed of two parts which originates in the galea aponeurotica, and extends posteriorly to reach the lateral areas of the occipital bone. Its contraction tenses the galea aponeurotica, which covers the skull.

greater zygomatic muscle
A long, thin facial muscle that goes from the cheek of the zygomatic bone to the skin of the commissure of the lips. When contracted, it elevates the commissure, complementing the action of the levator muscle of the angle of the mouth.

sternocleidomastoid muscle
A muscle which originates in the mastoid process of the temporal and occipital bones of the head. It descends forming two fascicles: one goes to the manubrium of the sternum and the other to the clavicle. Its action is to flex, lateralize and rotate the neck.

trapezius muscle
A very wide triangular muscle that covers almost all the other muscles of the nape of the neck and a large part of the back. It is inserted in the occipital bone and the spinous processes of the cervical and dorsal vertebrae, from where it converges on the shoulder where it is inserted in the scapula and the clavicle. Its action is to elevate the shoulder and to incline the head sideways.

27

NAPE OF THE NECK

▼ POSTERIOR VIEW

inferior oblique muscle of the head
Also known as the *greater oblique muscle*. It extends from the spinous process of the axis to the transverse process of the atlas. Its action is to rotate the head from one side to another.

rectus capitis posterior minor muscle
A muscle that extends from the atlas to the occipital bone and contributes to the inclination of the head backwards and sideways.

rectus capitis posterior major muscle
A flat muscle that unites the axis with the occipital bone. It assists the backwards and sideways inclination of the head and its rotation.

galea aponeurotica
A fibrous membrane that surrounds the superior area of the skull and is firmly joined to the skin covering it.

occipital muscle
A flat muscle that originates in the galea aponeurotica and extends to the lateral areas of the occipital bone. Its contraction tenses the skin of the skull.

semispinalis capitis muscle
A muscle that extends from the last cervical vertebrae and first dorsal vertebrae to the occipital bone. It can incline the head backwards or rotate it.

superior oblique muscle of the head
Also known as the *lesser oblique muscle*. It originates in the atlas and terminates in the occipital bone. It inclines the head laterally.

splenius muscle
A muscle located below the trapezius muscle which has a twin superior origin. One fascicle originates in the mastoid process of the temporal bone of the skull (splenius muscle of the head), and the other originates in the first cervical vertebrae (splenius muscle of the neck). Both fascicles descend and unite to be inserted in the last cervical vertebrae and the first dorsal vertebrae. They act to incline the head backwards or laterally or rotate it.

semispinalis capitis
The semispinalis capitis is a muscle that goes from the last cervical vertebrae to the mastoid process of the temporal bone. Its action is to incline the head backwards and to one side.

sternocleidomastoid muscle
A muscle which originates in the mastoid process of the temporal and occipital bones of the head. It descends, forming two fascicles: one goes to the manubrium of the sternum and the other to the clavicle. Its action is to flex, lateralize and rotate the neck.

longissimus muscle
An elongated muscle that unites the last cervical vertebrae and the first dorsal vertebrae with the mastoid process of the temporal bone. Its action is to incline the head backwards or to one side.

trapezius muscle
A very wide, triangular muscle that covers almost all the other muscles of the nape of the neck and a large part of the back. It is inserted in the occipital bone and the spinous processes of the cervical and dorsal vertebrae, where it converges on the shoulder and is inserted in the scapula and the clavicle. It serves to elevate the shoulder and to incline the head sideways.

superficial cervical fascia
A thin membrane or aponeurosis that covers all the structures of the neck, emitting individual prolongations that surround some muscles of the region.

28

NECK

▼ ANTERIOR VIEW

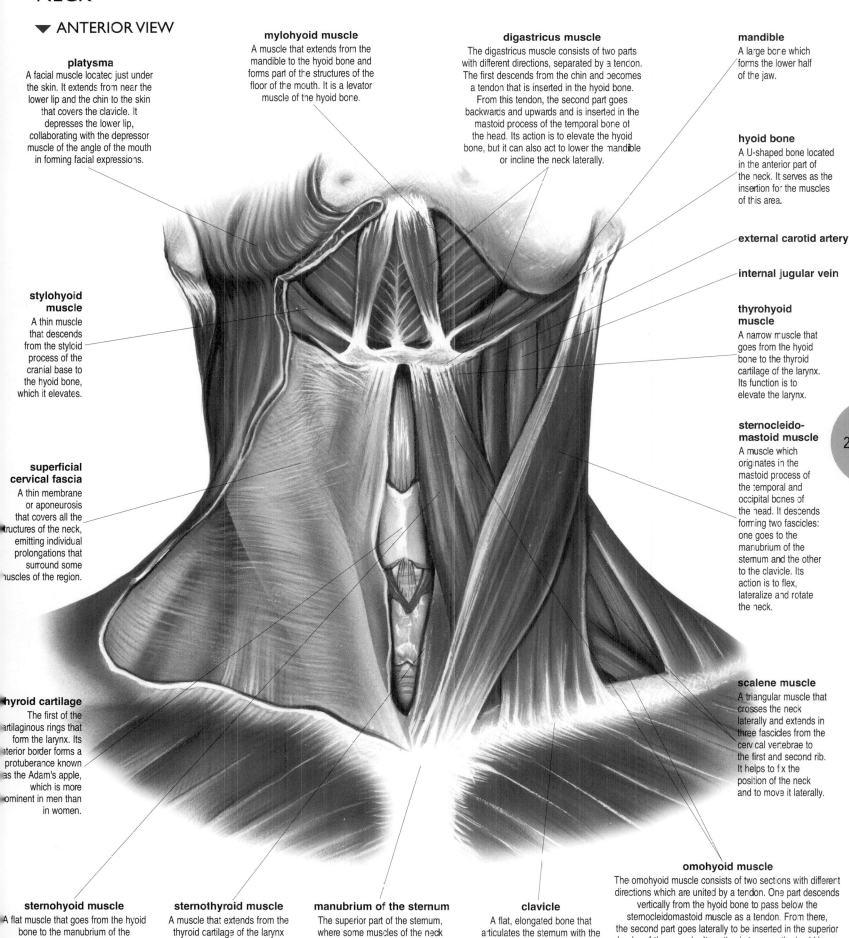

platysma
A facial muscle located just under the skin. It extends from near the lower lip and the chin to the skin that covers the clavicle. It depresses the lower lip, collaborating with the depressor muscle of the angle of the mouth in forming facial expressions.

mylohyoid muscle
A muscle that extends from the mandible to the hyoid bone and forms part of the structures of the floor of the mouth. It is a levator muscle of the hyoid bone.

digastricus muscle
The digastricus muscle consists of two parts with different directions, separated by a tendon. The first descends from the chin and becomes a tendon that is inserted in the hyoid bone. From this tendon, the second part goes backwards and upwards and is inserted in the mastoid process of the temporal bone of the head. Its action is to elevate the hyoid bone, but it can also act to lower the mandible or incline the neck laterally.

mandible
A large bone which forms the lower half of the jaw.

hyoid bone
A U-shaped bone located in the anterior part of the neck. It serves as the insertion for the muscles of this area.

external carotid artery

internal jugular vein

thyrohyoid muscle
A narrow muscle that goes from the hyoid bone to the thyroid cartilage of the larynx. Its function is to elevate the larynx.

stylohyoid muscle
A thin muscle that descends from the styloid process of the cranial base to the hyoid bone, which it elevates.

sternocleido-mastoid muscle
A muscle which originates in the mastoid process of the temporal and occipital bones of the head. It descends forming two fascicles: one goes to the manubrium of the sternum and the other to the clavicle. Its action is to flex, lateralize and rotate the neck.

superficial cervical fascia
A thin membrane or aponeurosis that covers all the structures of the neck, emitting individual prolongations that surround some muscles of the region.

scalene muscle
A triangular muscle that crosses the neck laterally and extends in three fascicles from the cervical vertebrae to the first and second rib. It helps to fix the position of the neck and to move it laterally.

thyroid cartilage
The first of the cartilaginous rings that form the larynx. Its anterior border forms a protuberance known as the Adam's apple, which is more prominent in men than in women.

sternohyoid muscle
A flat muscle that goes from the hyoid bone to the manubrium of the sternum. When contracted, it makes the hyoid bone and all the structures attached to it descend.

sternothyroid muscle
A muscle that extends from the thyroid cartilage to the manubrium of the sternum. Its contraction makes the larynx descend.

manubrium of the sternum
The superior part of the sternum, where some muscles of the neck are located.

clavicle
A flat, elongated bone that articulates the sternum with the scapula. It serves as the insertion point for many muscles of the neck and thorax.

omohyoid muscle
The omohyoid muscle consists of two sections with different directions which are united by a tendon. One part descends vertically from the hyoid bone to pass below the sternocleidomastoid muscle as a tendon. From there, the second part goes laterally to be inserted in the superior border of the scapula. Its action is to move the hyoid bone and its structures downwards and backwards.

THORAX

▼ ANTERIOR VIEW

pectoralis minor muscle
A flat muscle located below the pectoralis major muscle. It originates in the third, fourth and fifth ribs and ascends obliquely. Its insertion is in the coracoid process of the scapula. When contracted, it depresses the scapula and thus the entire shoulder. In addition it can elevate the ribs, becoming an inspiratory muscle.

subclavian muscle
A small muscle that ascends obliquely from the first costal cartilage to the inferior border of the clavicle. Its purpose is to make the clavicle and thus, the shoulder, descend.

pectoralis major muscle
A very wide triangular muscle, whose internal side originates in the anterior face of the sternum, the clavicle and the last ribs. It converges outwards and is attached by a tendon to the greater tubercle of the humerus. When contracted it lowers the arm when it is raised and, if it is already lowered, moves the shoulder forwards, bending the back. In addition it can elevate the thoracic cavity. It is innervated by the pectoral nerves.

intercostal muscles
Flat muscles located between the ribs, from the inferior border of the highest rib to the superior border of last rib. They consist of three layers of muscles; the internal, medial and external intercostals. Their purpose is to approximate the ribs to each other, widening or narrowing the chest as necessary during breathing.

serratus anterior muscle
A muscle located in the lateral wall of the thorax. It is formed of a series of fascicles that go from the nine or ten first ribs to the internal border of the scapula, bordering the thoracic wall laterally. When contracted it moves the internal border of the scapula forward, elevating the shoulder. In addition it has inspiratory functions, elevating the ribs and widening the thorax.

aponeurotic sheath of the rectus abdominis muscle
An aponeurotic sheath that covers the rectus abdominis muscle. Its internal border joins with the border of the contralateral muscle, forming the white line.

white line
A membrane which joins the aponeurotic sheaths that cover the superficial abdominal muscles. It marks the median line of the abdominal wall vertically.

rectus abdominis muscle
A flat muscle located in the anterior face of the abdomen, on either side of the white line. It originates in the costal cartilages of the fifth, sixth and seventh ribs and in the xiphoid appendix of the sternum. It descends vertically and is inserted in the superior border of the pubis. When contracted, it flexes the thorax forward or elevates the pelvis, while simultaneously compressing the abdominal organs. It plays an important role in defecation and childbirth.

external oblique muscle of the abdomen
A wide muscle located in the lateral wall of the abdomen. It originates in the lower ribs and extends obliquely in form of a fan formed by several fascicles that terminate in a tendinous membrane that merges with the sheath of the rectus abdominis muscle. Its contraction makes the ribs descend, flexes the thorax over the pelvis and inclines the thorax laterally.

30

THORAX

▼ POSTERIOR VIEW

deltoid muscle

A large muscle that occupies all the superficial area of the shoulder. It originates inserted in the clavicle and the scapula. It descends to become a tendon that is inserted in the external face of the humerus. Its action is to elevate the arm horizontally and also to move it backwards and forwards.

trapezius muscle

A very wide, triangular muscle that covers almost all the other muscles of the nape of the neck and a large part of the back. It originates in the occipital bone and the spinous processes of the cervical and dorsal vertebrae, from where it converges on the shoulder and it is inserted in the scapula and the clavicle. It elevates the shoulder and inclines the head sideways.

rhomboideus major muscle

A wide muscle that extends from the spinous processes of the first dorsal vertebrae to the internal border of the scapula. It acts to move the scapula inwards while inclining it.

rhomboideus minor muscle

A muscle located above the rhomboideus major muscle. It originates in the spinous processes of the last cervical vertebrae. It descends obliquely to the internal border of the scapula. When contracted, it tilts the scapula and thus depresses the shoulder.

levator scapulae muscle

A triangular muscle that originates in the transverse processes of the four or five first cervical vertebrae. It converges to be inserted in the medial border of the scapula. It acts to incline the scapula and depress the shoulder, also contributing to the lateral inclination of the head.

longissimus dorsi muscle

A muscle that originates in the common muscular mass of the erectors of the vertebral column. It ascends and is inserted in the transverse processes of the lumbar vertebrae and in the inferior border of the ribs. From there, it sends a prolongation to the last cervical vertebrae that is called the longissimus muscle of the neck. It is an extensor muscle of the vertebral column which, like the iliocostalis muscle, fixes and maintains the column erect.

supraspinatus muscle

A triangular muscle that originates in the supraspinous fossa of the posterior face of the scapula. It terminates in a tendon which is inserted in the trochlea of the head of the humerus. It acts as a levator for the arm and also contributes slightly to the internal rotation of the arm.

infraspinatus muscle

infraspinatus fascia

A membranous layer that covers the infraspinous muscle.

latissimus dorsi muscle

A very wide, thin muscle that extends across the lower back. The internal part originates in the spinous processes of the lumbar vertebrae and the dorsal vertebrae. The inferior part originates in the sacrum and the iliac crest and the superior part in the last three or four ribs. The muscle ascends towards the axilla and is inserted in the humerus through a tendon. With the arm raised, its contraction makes the humerus descend while rotating it internally. It also acts to elevate the ribs.

teres major muscle

A muscle that extends from the vertex and external border of the scapula to the humerus. Its function is to move the arm inwards and backwards. In addition, it tilts the scapula, acting as an elevating shoulder muscle.

serratus anterior muscle

A muscle located in the lateral wall of the thorax. It is formed of a series of fascicles that go from the nine or ten first ribs to the internal border of the scapula, laterally bordering the thoracic wall. When contracted, it moves the internal border of the scapula forwards, elevating the shoulder. In addition, it has inspiratory functions, elevating the ribs and widening the thorax

spinalis muscle

A muscle that originates in the erector spinae muscle group and, attached to the vertebral column, is inserted in the spinous processes of the lumbar and dorsal vertebrae. It continues towards the neck as the spinalis cervicis muscle. When contracted, it extends the vertebral column.

iliocostalis muscle

A long muscle that crosses all the back parallel to the vertebral column. It originates in the erector spinae muscle group and ascends to each of the ribs. It terminates in the transverse processes of the last cervical vertebrae. The iliocostalis muscle is an extensor muscle of the vertebral column that also inclines the column laterally while fixing and maintaining it upright.

serratus posterior (inferior) muscle

A quadrilateral-shaped muscle located below the wide dorsal muscle. It is inserted in the spinous processes of the last dorsal vertebrae and first lumbar vertebrae. It ascends as four stepped fascicles which are inserted in the inferior borders of the four last ribs. It is an inspiratory muscle that makes the last ribs descend and widens the thorax.

intercostal muscles

Flat muscles located between the ribs, from the inferior border of the highest rib to the superior border of last rib. They consist of three layers of muscles: the internal, medial and external intercostals. Their action is to approximate the ribs to each other, widening or narrowing the chest as necessary during breathing.

31

ABDOMEN

▼ ANTERIOR VIEW

serratus anterior muscle
A muscle located in the lateral wall of the thorax. It is formed of a series of fascicles that go from the nine or ten first ribs to the internal border of the scapula, bordering the thoracic wall laterally. When contracted it moves the internal border of the scapula forwards, elevating the shoulder. In addition, it has inspiratory functions, elevating the ribs and widening the thorax.

pectoralis major muscle
A very wide, triangular muscle, whose internal side originates in the anterior face of the sternum, the clavicle and the last ribs. It converges outwards and is inserted by a tendon in the greater tubercle of the humerus. When contracted it lowers the arm when it is raised and, if already it is lowered, moves the shoulder forwards, bending the back. In addition, it can elevate the thoracic cavity. It is innervated by the pectoral nerves.

external oblique muscle of the abdomen
A wide muscle located in the lateral wall of the abdomen. It originates in the last ribs and extends obliquely in form of a fan formed by several fascicles which terminate in a tendinous membrane that merges with the sheath of the rectus abdominis muscle. Its contraction makes the ribs descend, flexes the thorax over the pelvis and inclines the thorax laterally while compressing the organs of the abdominal cavity.

intercostal muscles
Flat muscles located between the ribs, from the inferior border of the highest rib to the superior border of the last rib. They consist of three layers of muscles: the internal, medial and external intercostals. Their action is to approximate the ribs to each other, widening or narrowing the chest as necessary during breathing.

32

rectus abdominis muscle
A flat muscle located in the anterior face of the abdomen, on either side of the white line. It originates in the costal cartilages of the fifth, sixth and seventh ribs and in the xiphoid appendix of the sternum. It descends vertically and is inserted in the superior border of the pubis. When contracted, it flexes the thorax forwards or elevates the pelvis, while simultaneously compressing the abdominal organs. It plays an important role in defecation and childbirth.

aponeurotic sheath of the rectus abdominis muscle
An aponeurotic sheath that covers the rectus abdominis muscle. Its internal border joins the border of the contralateral muscle to form the white line.

internal oblique muscle of the abdomen
A muscle located below the external oblique muscle. It originates in the anterosuperior iliac spine and the aponeurosis of the latissimus dorsi muscle and extends forwards like a fan. Superiorly, it is inserted in the cartilages of the last ribs, inferiorly in the pubis and it terminates medially in a wide membrane that merges with the sheath of the rectus abdominis muscle. It serves to lower the ribs, to flex the thorax or to incline it laterally and to compress the abdominal organs. Below this muscle, the transverse muscle of the abdomen follows a parallel path.

white line
A membrane that joins the aponeurotic sheaths that cover the superficial abdominal muscles. It marks the median line of the abdominal wall vertically, from the xiphoid process of the sternum to the pubis.

inguinal canal
A space located between the aponeurosis of the muscles of the inferointernal area of the abdomen. The inguinal canal is occupied by the spermatic cord in men and the round ligament in women.

pyramidalis muscle
A small, rudimentary muscle, whose function is not well-defined. It is located in the inferior part of the abdomen, in front of the rectus abdominis muscle. It originates in the superior border of the pubis and extends obliquely upwards to reach the white line.

spermatic cord
A cord-like organ containing the structures that connect with the testes, including the deferent duct, blood vessels and nerves.

ABDOMEN

▼ POSTERIOR VIEW

internal oblique muscle of the abdomen

A muscle located below the external oblique muscle. It originates in the anterosuperior iliac spine and the aponeurosis of the latissimus dorsi muscle. It extends forwards like a fan. Superiorly it is inserted in the cartilages of the last ribs, inferiorly in the pubis and medially, it terminates in a wide membrane that merges with the sheath of the rectus abdominis muscle. It serves to lower the ribs, to flex the thorax or to incline it laterally and to compress the abdominal organs.

erector spinae muscle group

A powerful muscular mass that originates in the spinal aponeurosis and iliac crest. It ascends in ramifications which include the iliocostalis muscle, the longissimus dorsi muscle and the spinalis muscle.

gluteus minimus muscle

A muscle located beneath the gluteus medius muscle. It originates in the anterior part of the iliac crest and the external iliac fossa and is inserted in the greater trochanter of the femur. Its function is similar to that of the gluteus medius muscle: separating and rotating the thigh.

piriformis muscle

A triangular muscle that originates in the anterior face of the sacrum and is inserted in the greater trochanter of the femur. It crosses the greater sciatic notch to leave the pelvis. Its contraction rotates the thigh outwards. When the thigh is flexed over the pelvis, as occurs in the seated position, the muscle abducts or separates the thigh.

dorsolumbar fascia

A thick, aponeurotic membrane that covers the muscles of the vertebral canal.

external oblique muscle of the abdomen

A wide muscle located in the lateral wall of the abdomen. It originates in the last ribs and extends obliquely in form of a fan formed by several fascicles that terminate in a tendinous membrane that merges with the sheath of the rectus abdominis muscle. Its contraction makes the ribs descend, flexes the thorax over the pelvis and inclines the thorax laterally while simultaneously compressing the organs of the abdominal cavity.

gluteus medius muscle

A very wide, thick muscle located below the gluteus maximus muscle. It originates in the iliac crest, the anterosuperior iliac spine, the external iliac fossa, the sacroiliac fibrous arch and the aponeurosis gluteus. It converges to be inserted in the greater trochanter of the femur. Its action is to raise the thigh in abduction or separation, while rotating it inwards and outwards. It is innervated by the superior gluteus nerve.

femoral aponeurosis or fascia lata

An aponeurotic sheath that covers and surrounds the muscles of the thigh. It extends from the pelvic area to the knee

gluteus maximus muscle

A thick muscle that corresponds to the buttocks. It originates in the iliac crest of the ilium, the sacrum, the coccyx and the lumbodorsal fascia. It descends obliquely as a large muscular mass inserted in the iliotibial tracts and the gluteal tuberosity presented by the femur below the greater trochanter. One part merges with the tensor muscle of the fascia lata. It serves to extend the thigh backwards while rotating it outwards. It also helps to maintain the body upright by fixing the pelvis over the femur.

superior gemellus muscle

A flat muscle that originates in the sciatic spine of the iliac bone and goes outwards horizontally from there. It merges with the internal obturator muscle and the inferior gemellus muscle in a terminal tendon that is inserted in the greater trochanter of the femur. Its function is to rotate the thigh outwards.

inferior gemellus muscle

A flat muscle that originates in the tuberosity of the ischium and extends outwards to merge with the superior gemellus and the internal obturator muscles, forming a terminal tendon that is inserted in the greater trochanter of the femur. Its action is similar to that of the two muscles to which it is united, rotating the thigh outwards.

quadratus femoris muscle

A square muscle located in the posterior part of the hip joint. It originates in the tuberosity of the ischium and is inserted in the posterior border of the femur. Its function is to rotate the thigh outwards.

internal obturator muscle

A muscle that follows a passage parallel to the two geminus muscles which it separates. It originates in the obturator membrane that covers the obturator foramen of the pelvis and in the bone ischium and pubis. It terminates with the superior and inferior gemellus muscles in a common tendon which is inserted in the greater trochanter of the femur. When contracted, it rotates the thigh outwards.

spinal aponeurosis

A strong, pearly-colored rhomboid membrane attached to the iliac crests and the sacrum and which serves as the inferior insertion point for the iliocostalis and longissimus dorsi muscles.

DIAPHRAGM

▼ SUPERIOR VIEW

lumbar vertebra
The first lumbar vertebrae serve as the insertion points for the tendons which terminate the posterior part of the diaphragm muscle and its transverse processes.

spinal marrow
Part of the nervous system which runs down the spine and is protected by the vertebrae. It is the origin of the spinal nerves.

intervertebral disc
A cartilaginous disc located between the bodies of the vertebrae. Its function is to cushion the messages that are transmitted from one vertebra to another.

pleura
A membrane covering the lungs which are attached to the cupola of the diaphragm by their anterior faces.

azygos vein
A vein that ascends together with the vertebral bodies and joins the superior vena cava in the superior part of the thorax. In its trajectory, it collects the blood coming from the intercostal veins.

intercostal muscles
Flat muscles located between the ribs, from the inferior border of the highest rib to the superior border of last rib. They consist of three layers of muscles: the internal, medial and external intercostals. Their action is to approximate the ribs to each other, widening or narrowing the chest as necessary during breathing.

DIAPHRAGM MUSCLE
A flat muscle that separates the thoracic and abdominal cavities. It is shaped like a concave vault. The diaphragm muscle is inserted posteriorly in the first lumbar vertebrae and the last ribs and anteriorly in the xiphoid process of the sternum and the last ribs.

thoracic aorta
A large, thick blood vessel that originates in the heart and crosses the thorax vertically, with branches that irrigate the thoracic organs. When it crosses the diaphragm to the abdomen, it becomes the *abdominal aorta*.

esophagus
A tubular duct forming part of the alimentary canal. It links the pharynx with the stomach after crossing the diaphragm.

pericardium
A saccular membrane which covers the heart and adheres to the cupola of the diaphragm in its inferior face.

ribs
The last ribs serve as the insertion points for the diaphragmatic muscle and some muscles of the abdominal walls.

inferior vena cava
A thick venous duct that collects the blood from the inferior extremities and the abdomen and carries it to the heart. As it ascends, it crosses the diaphragm.

sternum
A flat bone located in the anterior face of the thorax, in whose lateral borders the ribs unite to close the thoracic cavity.

DIAPHRAGM

▼ INFERIOR VIEW

hiatus of the vena cava
An orifice located near the phrenic center, through which the inferior vena cava passes from the abdomen to the thorax.

esophageal hiatus
An orifice located in the center of the diaphragm, through which the esophagus passes from the thorax to the abdomen.

aortic hiatus
An orifice located under the median arcuate ligament, between the crus, through which the aorta passes from the thorax to the abdomen

DIAPHRAGM MUSCLE
A flat muscle that separates the thoracic and abdominal cavities. It is shaped like a concave vault. The diaphragm muscle is inserted posteriorly in the first lumbar vertebrae and the last ribs and anteriorly in the xiphoid process of the sternum and the last ribs.

median arcuate ligament
A ligament that unites the crus of the diaphragm, leaving an orifice through which the aorta passes from the thorax to the abdomen.

crus of the diaphragm
The left and right crus are formed of muscle and tendon and are located in the posterior part of the diaphragm which they attach to the bodies of the first lumbar vertebrae.

lateral arcuate ligament
A ligament in the form of an arch which unites the transverse process of the first lumbar vertebra with the twelfth rib. The muscle lumbar passes under the ligament.

quadratus lumborum muscle
A flat muscle that extends vertically over the posterior face of the abdomen, from the last rib to the iliac crest, emitting fascicles that are inserted in the transverse processes of the lumbar vertebrae. Its action is to incline the vertebral column laterally.

psoas major muscle
A long muscle that crosses all the posterior face of the abdomen. It originates in the last rib, leaving the pelvic cavity to be inserted in the smaller trochanter of the femur. When contracted, it flexes the thigh and inclines the vertebral column forwards or laterally.

medial arcuate ligament
A ligament that forms an arch between the posterior pillar of the diaphragm and the transverse process of the first lumbar vertebra. The psoas major muscle passes under the ligament.

transverse muscle of the abdomen
A muscle that crosses the lateral wall of the abdomen transversally, from the transverse processes of the lumbar vertebrae and the iliac crest to the anterior face, where it merges with a wide membrane that covers the other muscles of the anterior face.

MALE PERINEUM

bulbocavernosus muscle
An erector muscle that originates in the prerectal area and goes upwards and forwards, bordering the spongy portion of the urethra and terminating in the cavernous bodies of the penis.

penis
The male genital organ. It contains internal cavernous bodies which fill with blood during sexual arousal and makes the penis erect. The penis carries the male urethra, through which both semen and urine are expelled.

ischiocavernosus muscle
A muscle that goes from the cavernous bodies of the penis to the ischium. It facilitates the entrance of blood in the penile cavernous bodies and to provoke the erection of the penis.

deep fascia of the penis
A cylindrical, membranous sheath that surrounds the cavernous bodies of the penis.

superficial transverse muscle of perineum
A small muscle that goes from the bone ischium towards the median line, and is inserted in the prerectal raphe. The deep transverse muscle of the perineum follows a parallel course. Its action is complementary to that of the levator muscle of the anus, contributing to the process of defecation. It also intervenes in urination and ejaculation.

prerectal raphe
A membrane that unites the anterior part of the anus with the base of the penis.

36

ischiococcygeus muscle
A flat, triangular muscle located behind the levator muscle of the anus. It is inserted anteriorly in the sciatic spine and extends to the coccyx. It acts to support the intrapelvic organs.

anus
The external opening of the rectum through which the fecal matter is expelled.

gluteus maximus muscle
A thick muscle that corresponds to the buttocks. It extends the iliac crest, the sacrum, the coccyx and the aponeurosis and ligaments of the area, to the femur. It acts to extend the thigh backwards, to rotate it outwards and to fix the pelvis in the upright position.

termination of the coccyx
The last coccygeal vertebrae, covered by an aponeureosis and serving as the insertion point for various muscles of the region.

external anal sphincter
A ring-shaped muscle which surrounds the anal orifice and has prolongations to the skin of the perineum and the anococcygeal raphe. It acts as a sphincter, impeding defecation when contracted and relaxing to allow fecal matter to pass.

levator muscle of the anus
A flat muscle that goes from the pubic bone to the rectum and the anus, passing by the lateral wall of the prostate. It consists of two fascicles, one superficial and another deep. It contributes to the act of defecation by compressing the rectum and elevating the anus, and also supports the intrapelvic organs.

FEMALE PERINEUM

gracilis muscle
A muscle of the leg, not the perineum. It crosses the internal part of the thigh, from the pubis to the femur. It flexes the thigh.

ischiocavernosus muscle
A muscle that goes from the pelvis to the base of the clitoris. It facilitates the entrance of blood to the cavernous bodies of the clitoris and to provoke its erection.

clitoris
An erectile organ located in the vertex of union of the two labia majora. It is formed of cavernous tissue that fills with blood during sexual arousal.

urethral orifice
A small orifice located below the clitoris and above the vaginal orifice (the termination of the urethra) where urine is expelled.

bulbocavernosus muscle
A muscle that originates in the anterior region of the anus and extends forwards to the distal area of the vagina and the urethra, reaching the base of the clitoris. It acts to facilitate the erection of the clitoris and vaginal contraction and the secretion of the mucous glands during intercourse.

labia minora
Mucocutaneous folds that border the vaginal orifice laterally.

vaginal orifice
An orifice which communicates the vagina and the vulva and which is located below the urethral ostium. It receives the penis during intercourse and acts as the birth canal.

superficial transverse muscle of perineum
A small muscle that goes from the ischium to the prerectal raphe. Its action is complementary to that of the levator muscle of the anus, contributing to the process of defecation.

perineal raphe
A membrane that unites the posterior part of the vaginal orifice with the anterior part of the anus, below the perineum.

ischiococcygeus muscle
A flat, triangular muscle located behind the levator muscle of the anus. It is inserted anteriorly in the sciatic spine and extends to the coccyx. It supports the intrapelvic organs.

anus
The external opening of the rectum through which the fecal matter is expelled.

gluteus maximus muscle
The gluteus maximus is a muscle of the pelvis not the perineum. It goes from the coccyx, the sacrum and the iliac crests to the femur. It is essential for walking and for maintaining the stability of the pelvis.

levator muscle of the anus
A flat muscle that goes from the pubic bone to the rectum and the anus. It consists of two fascicles, one superficial and another deep. It contributes to the act of defecation by compressing the rectum and elevating the anus. It also supports the intrapelvic organs.

external anal sphincter
A ring-shaped muscle which surrounds the anal orifice. It acts as a sphincter, impeding defecation when contracted and relaxing to allow it.

termination of the coccyx
The last coccygeal vertebrae which are covered by an aponeurosis and serve as the insertion for various muscles of the region.

anococcygeal ligament
A membrane that extends from the posterior border of the anus to the last vertebrae of the coccyx.

37

SHOULDER AND ARM: SUPERFICIAL MUSCLES

▼ ANTERIOR VIEW

pectoralis major muscle
A very wide, triangular muscle, whose internal side is inserted in the anterior face of the sternum, the clavicle and the last ribs. It converges outwards and is inserted by a tendon in the greater tubercle of the humerus. When contracted, it lowers the arm when it is raised and, if already it is lowered, moves the shoulder forwards, bending the back. In addition, it can elevate the thoracic cavity. It is innervated by the pectoral nerves.

clavicle
An elongated bone that joins the sternum with the scapula. It is the insertion for various muscles of the neck, shoulder and pectoral area.

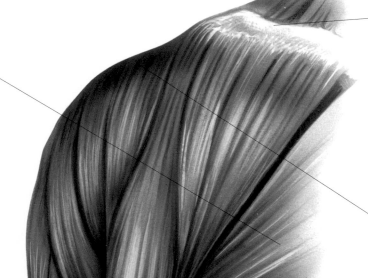

deltoid muscle
A voluminous muscle that occupies the entire superficial area of the shoulder. It is inserted in the clavicle and the scapula. It descends, becoming a tendon that is inserted in the external face of the humerus. Its action is to elevate the arm to the horizontal and also to move it backwards and forwards.

38

biceps brachii muscle
A thick muscle that occupies the anterior face of the arm. It consists of two parts: an external or long portion which originates in the external angle of the scapula and a short, internal portion which originates in the coracoid process of the scapula. The two portions merge to form a single muscular mass that becomes a tendon which crosses the elbow and is inserted in the head of radius. The biceps muscle flexes the forearm on the arm, places it in supination, or external rotation, and elevates the arm.

extensor carpi radialis longus
A flat muscle located below the brachiradialis muscle. It originates in the external border of the humerus, crosses the external border of the forearm and terminates as a tendon that crosses the wrist joint and is inserted in the base of the second metacarpal bone of the hand. When contracted, it extends the second metacarpal bone, moving the hand in extension over the forearm.

brachialis muscle
A very wide muscle located below the biceps muscle. It originates in the internal and external faces of the humerus, from where it descends, crossing the anterior part of the elbow, to be inserted in the ulna. Its main action is to bend the forearm over the arm.

epitrochlea or internal condyle
A bony protuberance located in the internal area of the inferior extremity of the humerus, where ligaments of the elbow joint and muscles of the forearm are inserted.

brachioradialis muscle
A muscle which originates in the external border of the humerus and, after crossing all the forearm, is converted in a tendon which is inserted in the inferior end of radius. It flexes the forearm.

SHOULDER AND ARM: SUPERFICIAL MUSCLES

▼ POSTERIOR VIEW

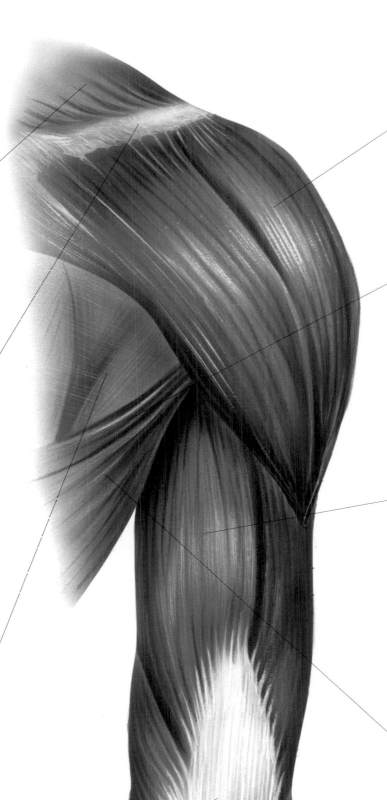

trapezius muscle
A very wide, triangular muscle that covers almost all the other muscles of the nape of the neck and a large part of the back. It originates in the external protuberance of the occipital bone and the spinous processes of the seven cervical and twelve dorsal vertebrae. The vertex of the triangle is located in the shoulder where the trapezius is inserted in the acromion and spine of the scapula and the clavicle. It elevates the shoulder and inclines the head sideways.

spine of the scapula
An elevated protuberance located in the posterior face of the scapula, between the deltoid and trapezius muscles.

infraspinatus fascia
A membranous layer that covers the infraspinous muscle. It occupies almost all the posterior face of the scapula and extends to the head of the humerus where it is inserted.

tendon of triceps brachii
A strong, muscular mass resulting from the union of the three parts that form the brachial triceps. Becomes a tendon that reaches the elbow and is inserted in the olecranon of the ulna.

deltoid muscle
A voluminous muscle that occupies the entire superficial area of the shoulder. It is inserted in the clavicle and the scapula. It descends to become a tendon that is inserted in the external face of the humerus. Its action elevates the arm horizontly and also moves it backwards and forwards.

teres major muscle
A muscle that extends from the vertex and external border of the scapula to the humerus. It moves the arm inwards and backwards. In addition, it tilts the scapula, acting as an elevating muscle for the shoulder

triceps brachii muscle
A thick muscle that occupies the posterior area of the arm. Its superior part consists of three portions: the long portion, which originates in the external border of the scapula, the external portion, or vastus externus, which originates in the posterior face of the humerus and the internal portion, or vastus medialis muscle, which originates in the posterointernal face of the humerus. The three portions unite to form a thick muscular mass which terminates in a tendon which is inserted in the olecranon of the ulna. The triceps brachii is an extensor of the forearm over the arm.

latissimus dorsi muscle
A very wide, thin muscle that extends across the inferior area of the back. The internal part is inserted in the spinous processes of the lumbar vertebrae and the dorsal vertebrae. The inferior part is inserted in the sacrum and the iliac crest and the superior part in the last three or four ribs. The muscle ascends towards the axilla and is inserted through a tendon in the humerus. With the arm raised, its contraction makes the humerus descend while rotating it internally. It also acts to elevate the ribs.

brachioradialis muscle
A long muscle which extends along the external border of the forearm. It originates in the external border of the humerus and, after crossing all the forearm, is converted into a tendon that is inserted in the inferior end of radius. Its main action is to flex the forearm over the arm, although it also rotates the forearm outwards when it is rotated inwards, or inwards when it is rotated outwards.

39

FOREARM: SUPERFICIAL MUSCLES

▼ ANTERIOR VIEW

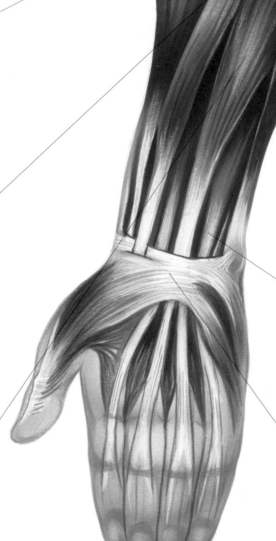

tendon of the biceps brachii muscle

brachioradialis muscle
A long muscle which extends along the external border of the forearm. It originates in the external border of the humerus and, after crossing the forearm, is converted into a tendon inserted in the inferior end of radius. Its main action flexes the forearm over the arm, although it also rotates the forearm outwards when it is rotated inwards, or inwards when it is rotated outwards.

pronator teres muscle
A flat muscle extending obliquely from the epitrochlea of the humerus and the coronoid process of the ulna to the external face of the radius. It rotates the forearms inwards in pronation and also flexes the forearm over the arm.

flexor carpi radialis muscle
A muscle that crosses the anterior face of the forearm obliquely, from the epitrochlea of the humerus to the second metacarpal bone of the hand. When contracted, it flexes the hand on the forearm and the forearm on the arm. It also moves the hand outwards.

palmaris longus muscle
The flexor carpi radialis muscle follows a parallel passage to the greater annular ligament of the wrist and the palmar aponeurosis of the hand, where it is inserted by a long tendon. It flexes the hand anteriorly on the forearm and tenses the palmar aponeurosis.

epitrochlea or internal condyle
A bony protuberance located in the internal area of the inferior extremity of the humerus, where ligaments of the elbow joint and muscles of the forearm are inserted.

aponeurotic sheath of the forearm
A cylindrical sheath that totally covers the muscles of the forearm and the arm. In its internal superior part it is inserted in the epitrochlea of the humerus.

flexor carpi ulnaris muscle
A muscle that occupies the internal border of the forearm. It originates in the superior part of the epitrochlea of the humerus and in the olecranon of the ulna. It descends the internal border of the ulna, crosses the wrist joint and is inserted by a tendon in the carpus. It flexes the hand on the forearm, turning the palm outwards.

flexor retinaculum
A fibrous ligament that covers the anterior face of the carpus. It serves as the insertion for many muscles of the palm of the hand. The tendons of the muscles flexors pass below the ligament to the palmar region.

flexor digitorum superficialis muscle
A wide muscle that occupies almost all the medial plane of the anterior face of the forearm. It originates in the epitrochlea of the humerus and the coronoid process of the ulna. It forms a wide muscular mass which divides into four muscular fascicles terminating in four tendons that cross the wrist joint below the annular ligament of the carpus. In the palm of the hand, they go to the second phalanges of the second, third, fourth and fifth fingers. It flexes the second phalange of the fingers toward the first, the fingers toward the hand, the hand toward the forearm and the forearm toward the arm.

FOREARM: SUPERFICIAL MUSCLES

▼ POSTERIOR VIEW

epitrochlea or internal condyle
A bony protuberance located in the internal area of the inferior extremity of the humerus, where ligaments of the elbow joint and muscles of the forearm are inserted.

anconeus muscle
A flat, triangular muscle which originates in the epicondylus of the humerus and which extends to be inserted in the posterior border of the ulna. It collaborates with the brachial triceps muscle, extending the forearm over the arm.

flexor carpi ulnaris muscle
A muscle that occupies the internal border of the forearm. It originates in the superior part of the epitrochlea of the humerus and in the olecranon of the ulna. It descends the internal border of the ulna, crosses the wrist joint and is inserted by a tendon in the carpus. It flexes the hand on the forearm, turning the palm outwards.

aponeurotic sheath of the forearm
A cylindrical sheath that totally covers the muscles of the forearm and the arm. It is inserted, in its internal superior part, in the epitrochlea of the humerus.

extensor carpi ulnaris muscle
A muscle that extends obliquely across the posterior area of the forearm, from the epicondylus of the humerus, and crosses the wrist joint, terminating in the fifth metacarpal bone. When contracted, it doubles the hand backwards over the forearm in extension, while simultaneously inclining it inwards in adduction.

extensor digiti minimi
A long, thin muscle that crosses the posterior area of the forearm, from the humeral epicondylus to the two last phalanges of the fifth finger, where it is inserted by means of a tendon that merges in its terminal part with the tendon of the extensor digitorum muscle. It acts to extend the fifth finger.

extensor retinaculum
A fibrous ligament that extends along the posterior face of the wrist joint. The tendons of the muscles of the posterior area of the forearm pass under this ligament to the hand.

tendon of the triceps brachii muscle

extensor carpi radialis longus
A flat muscle located below the brachiradialis muscle. It originates in the external border of the humerus, crosses the external border of the forearm and terminates as a tendon that crosses the wrist joint and is inserted in the base of the second metacarpal bone of the hand. When contracted, it extends the second metacarpal bone, moving the hand in extension over the forearm.

extensor digitorum muscle
A flat muscle that originates in the epicondylus of the humerus. It descends and divides into three fascicles which, converted into tendons, cross under the extensor retinaculum and terminate in the second and third phalanges of the second, third, fourth and fifth fingers. It acts to extend the third phalange over the second, the second over the first, the fingers over the hands, the hand over the forearm and the forearm over the arm.

extensor carpi radialis brevis
A muscle located in the external area of the forearm, below the first external radial. It extends from the epicondylus of the humerus and the external lateral ligament of the elbow joint to the wrist joint and is inserted in the base of the third metacarpal bone. When contracted, it turns the hand backwards over the forearm in extension.

abductor pollicis longus
A muscle that extends from the posterior face of the ulna and radius to the base of the first metacarpal bone, crossing the wrist joint under the posterior annular ligament. It turns the thumb outwards in abduction, as well as the rest of the hand, and rotates it.

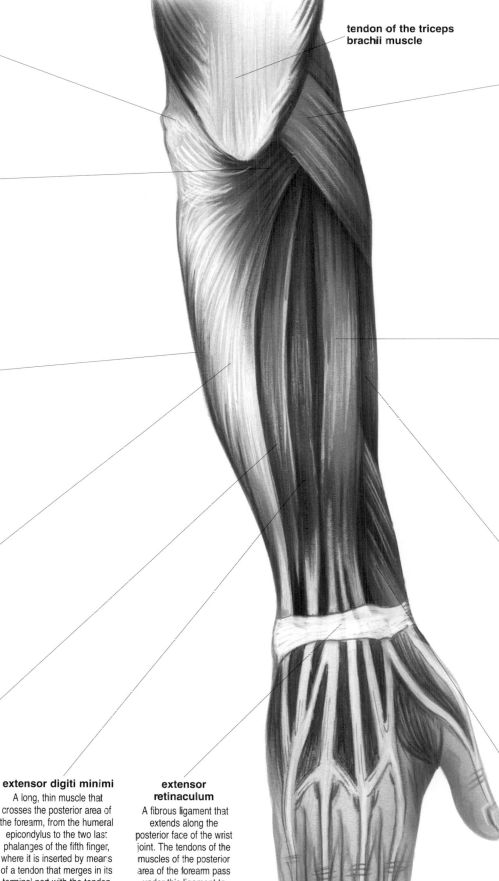

HAND: SUPERFICIAL MUSCLES

▼ ANTERIOR VIEW

opponens digiti minimi muscle

A muscle located below the short flexor and the separator of the fifth finger. It originates in the flexor retinaculum and is inserted in the fifth metacarpal bone. It moves the fifth finger forwards and towards the median line of the hand in opposition to the thumb.

palmar interosseal muscles

Small muscles located in the palmar face between the metacarpal bones. From there, they descend to merge with the corresponding tendon of the extensor digitorum muscle of the second, fourth and fifth fingers. Their action is to flex the first phalanges and to extend the second and third, at the same time approximating the last four fingers to each other. They are innervated by branches of the ulnar nerve.

tendons of the flexor digitorum profundis muscle

The tendons of this muscle, whose muscular mass is found in the deep area of the anterior face of the forearm, when they reach the height of first phalanges of the four last fingers, pass through a small opening created by the bifurcation of the tendons of the flexor digitorum superficialis muscle. They terminate in the third phalanges of the four last fingers. They act to flex the fingers over the hand and the hand over the forearm.

tendons of the flexor digitorum superficialis muscle

The tendons of the muscle divide in two at the height of the first phalange of the four last fingers, leaving small openings through which the tendons of the deep common flexor muscle pass. The two tendons are inserted in the lateral faces of the second phalanges. They flex the fingers over the hand and the hand over the arm.

flexor digiti minimi muscle

A muscle that follows a parallel path to the abductor digiti minimi muscle. It extends from the carpus and the flexor retinaculum to the base of the first phalange of the fifth finger, where it is inserted by a tendon common to the abductor digiti minimi muscle. Its action doubles the first phalange of the fifth finger over the palm of the hand.

adductor pollicis muscle

A triangular-shaped muscle which consists of two fascicles: one oblique, originating in the carpal bones, and the other transverse, originating in the second and third metacarpal bones. The muscle converges and is inserted in the first phalange of the thumb. It carries the thumb inwards, in a movement of approach or adduction.

flexor pollicis brevis muscle

A muscle located below the abductor or short separator of the thumb. It originates in the flexor retinaculum and the carpal bones and is inserted in the first phalange of the thumb. Its action is to move the thumb forwards and inwards.

abductor digiti minimi

A muscle which occupies the internal border of the palm of the hand, from the carpus to the first phalange of the fifth finger. When contracted, it separates the fifth finger from the central axis of the hand and flexes the first phalange of the fifth finger over the palm of the hand.

abductor pollicis brevis muscle

A muscle, located in the superficial area of the thenar eminence, which goes from the flexor retinaculum to the first phalange of the thumb. When contracted, it separates or abducts the thumb outwards, while simultaneously moving it forwards.

flexor retinaculum

A fibrous ligament that covers the anterior face of the carpus. It serves as the insertion for many of the muscles in the palm of the hand. The tendons of the muscle flexors pass below the ligament to the palmar region.

opponens pollicis muscle

A small, triangular muscle which originates in the flexor retinaculum and goes to the first metacarpal bone. When contracted, it carries the first metacarpal bone and the thumb forwards and inwards, and while rotating it slightly internally, leaves the thumb in opposition to the other four fingers.

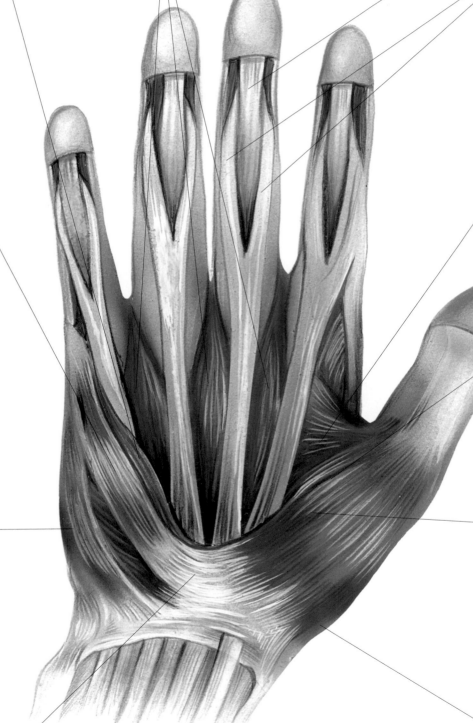

hypothenar eminence

thenar eminence

42

HAND: SUPERFICIAL MUSCLES

▼ POSTERIOR VIEW

tendons of the extensor digitorum muscle

These muscle tendons are located in the posterior face of the forearm. They reach the three phalanges of the second, third fourth and fifth fingers, where they are inserted. They act to extend the phalanges, the fingers and the hand.

tendons of the flexor digitorum superficialis muscle

The muscle tendons divide into two at the height of the first phalange of the last four fingers, leaving small openings through which the tendons of the deep common flexor muscle pass. The two tendons are inserted in the lateral faces of the second phalanges. They act to flex the fingers toward the hand and the hand toward the arm.

dorsal interosseal muscles

Small muscles located on the dorsal surface of the hand in the spaces between the metacarpal bones. They descend to merge with the corresponding tendon of the extensor digitorum muscle of the second, fourth and fifth fingers. They act to flex the first phalanges and to extend the second and third phalanges while simultaneously separating the last four fingers.

tendon of the extensor pollicis brevis muscle

This muscle tendon, whose mass is located in the posterior face of the forearm, reaches the first phalange of the thumb and extends it.

tendon of the extensor pollicis longus muscle

This muscle tendon is located deep within the posterior face of the forearm. It reaches the second phalange of the thumb, serving to extend it.

tendon of the extensor digiti minimi

The tendon of this muscle, whose muscular mass is in the posterior face of the forearm, merges with the extensor digitorum muscle in a tendon which goes to the fifth finger. It serves to extend the fifth finger.

extensor retinaculum

A fibrous ligament that extends along the posterior face of the carpus. The tendons of the extensor muscles pass under this ligament to the dorsal face of the hand.

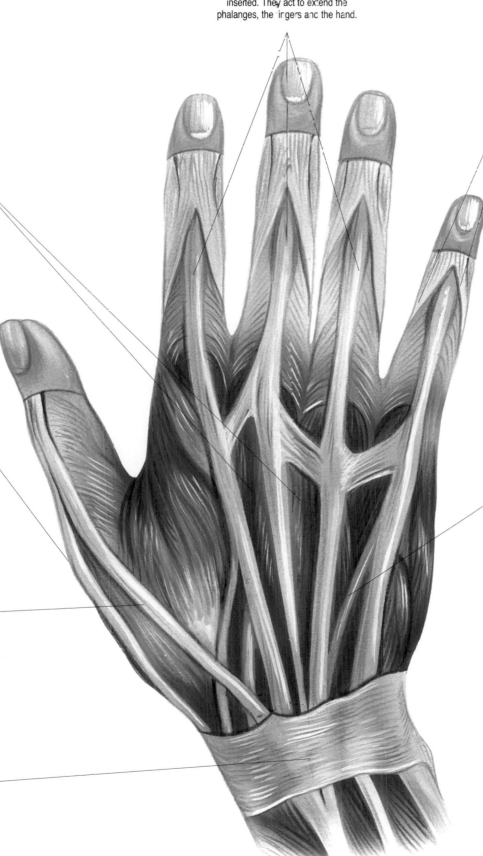

43

THIGH: SUPERFICIAL MUSCLES

▼ ANTERIOR VIEW

anterosuperior iliac spine
A bony protuberance which is the anterior limit of the iliac crest.

iliopsoas muscle
A muscle consisting of two parts: the psoas major, which originates in the last rib and the lumbar vertebrae, and the iliac which originates in the ilium and the sacrum. The two parts join and when they reach the thigh they are inserted in the smaller trochanter of the femur. Its action is to double the thigh over the pelvis, to incline it forwards and to fix it to the column in the upright position.

pectinate muscle
A flat muscle that joins the pelvis and the thigh. It originates in the pubic bone and the inguinal ligament, and descends obliquely to be inserted as a tendon in the smaller trochanter of the femur. When contracted, it approximates the thigh inwards in adduction while simultaneously rotating the thigh outwards. It also collaborates in the flexion of the thigh over the pelvis.

tensor muscle of the fascia lata
A flat muscle that originates in the iliac crest, anterosuperior iliac spine and the aponeurosis of the gluteus muscles. It descends laterally with a fascicle that merges with the ligaments of the fascia lata or femoral aponeurosis, and other fascicles that terminate in the external tuberosity of the tibia. When contracted, it stretches the external part of the femoral aponeurosis, while simultaneously abducting or separating the thigh. It also helps to incline the pelvis to one side and fix it in the standing position over the inferior extremities.

pubic bone
The anterior bone of the three that form the hip bone. In this area it presents a crest or pectinate line, in which muscles of the thigh and the pelvis are inserted.

gracilis muscle
A long, thin muscle that originates in the pubic bone and follows the internal border of the thigh to reach the superior part of the tibia, where it is inserted by a tendon that is common to two other muscles, the semitendinosus and the sartorius. The tendon is known as the pes anserinus. The gracilis muscle is a flexor of the leg over the thigh and an approximator of the thigh.

sartorius muscle
A long, thin muscle that crosses the anterior face of the thigh obliqely, from the anterosuperior iliac spine to the internal part of the superior extremity of the tibia. In this area, it shares a thick tendon with the gracilis muscle and semitendinosus muscle, a tendon that goes by the name of pes anserinus. When contracted, it doubles the leg on the thigh and the thigh on the pelvis, while simultaneously rotating and separating the thigh externally.

adductor longus muscle
Also called the first adductor muscle. It extends from the pelvis to the thigh and follows a parallel passage to the pectinate muscle. It originates in the pubic bone and is inserted in the internal border of the femur. Like the other adductor muscles, it approximates the thigh inwards and rotates it outwards.

rectus femoris muscle
A portion of the quadriceps that occupies the central position between two tendons in the anteroinferior iliac spine and the articular capsule of the hip joint.

quadriceps femoris muscle
A thick muscle that occupies the anterior face of the thigh. It is formed of four fascicles: the vastus lateralis muscle, vastus medialis muscle, the rectus femoris muscle and the vastus intermedius muscle, which is in a deeper plane. These fascicles finish in a wide common tendinous aponeurosis, which is inserted in the patella and then descends as the patellar tendon, terminating in the anterior tuberosity of the tibia. It serves to extend the leg over the thigh, although it also acts to double the thigh over the pelvis.

vastus lateralis muscle
The external portion of the four muscles that compose the quadriceps. It originates in the greater trochanter of the femur.

patellar ligament
A thick ligament that extends from the vertex of the patella to the anterior tuberosity of the tibia. It is a prolongation of the tendon of inferior insertion of the four portions of the quadriceps muscle.

patella
A flat, round bone that occupies the anterior face of the knee joint.

vastus medialis muscle
A part of the quadricep muscle which is attached to the internal part of the femur. It originates in the area of transition between the body and the neck of this bone.

44

THIGH: SUPERFICIAL MUSCLES

▼ POSTERIOR VIEW

gluteus maximus muscle

A thick muscle that corresponds to the buttocks. It originates in the iliac crest of the ilium, the sacrum, the coccyx and the lumbodorsal fascia. It descends obliquely as a large muscular mass inserted in the iliotibial tracts and the gluteal tuberosity presented by the femur below the greater trochanter. One part merges with the tensor muscle of the fascia lata. Its main action is to extend the thigh backwards while rotating it outwards. It also helps to maintain the body upright by fixing the pelvis over the femur.

adductor magnus muscle

A wide muscle that extends from the ischium of the pelvis to the internal border of the femur. It is located below the sartorius, adductor minimis and quadriceps muscles of the anterior face of the thigh. Its action is to approximate the thigh inwards and to rotate it.

gracilis muscle

A long, thin muscle that originates in the pubic bone and follows the internal border of the thigh to reach the superior part of the tibia, where it is inserted by a tendon common to two other muscles, the semitendinosus and the sartorius. The tendon is known as the pes anserinus. The gracilis muscle is a flexor of the leg over the thigh and an approximator of the thigh.

semitendinosus muscle

A muscle that originates in the ischium and descends the posterior face of the thigh. After passing the internal border of the knee, it becomes anterior and is converted into the tendon known as the pes anserinus, which serves as the common insertion in the superior extremity of the tibia for two other muscles: the sartorius and the gracilis muscles. When contracted, it doubles the leg over the thigh and rotates the thigh internally. It also acts as an extensor of the thigh over the pelvis.

thoracolumbar aponeurosis

A thick membrane that covers the musculature of the back and reaches the iliac crest of the pelvis.

semimembranous muscle

A muscle partially covered by the semitendinosus muscle. It is called the semimabranous muscle because its superior third is composed of a wide tendinous membrane. It originates in the ischium and is inserted in the superior extremity of the tibia. Its action is similar to the semitendinosus muscle.

gluteal aponeurosis

An extremely thick aponeurotic membrane that covers the gluteus musculature externally. It originates in the iliac crest and terminates in the thigh where it merges with the femoral fascia or fascia lata.

short head of the biceps femoris muscle

The short portion of the two that form the superior part of the biceps femoral muscle. It originates in the internal border of the femur, in the linea aspera.

iliotibial tract

A fibrous membrane that crosses the thigh laterally and superficially and is formed by the prolongation of the membrane that covers the thigh called the fascia lata.

biceps femoris muscle

A thick muscle that crosses the external part of the dorsal region of the thigh. The superior part is formed of two portions, the short and long portions, which merge and terminate in a single, very long tendon that is inserted in the superior extremity of the fibula. The biceps femoral flexes the leg over the thigh while simultaneously rotating the thigh slightly externally. It also acts as an extensor of the thigh over the pelvis.

long head of biceps femoris muscle

One of the two portions forming the biceps femoris muscle in its superior part. It originates in the ischium and descends obliquely down the posterior face of the thigh.

popliteal fossa

A rhomboidal space located in the dorsal face of the knee, framed by the muscles of the posterior face of the thigh and the leg (semimembranous, biceps femoris and gastrocnemius muscles). It is a passage for the blood vessels and the nerves from the thigh to the leg.

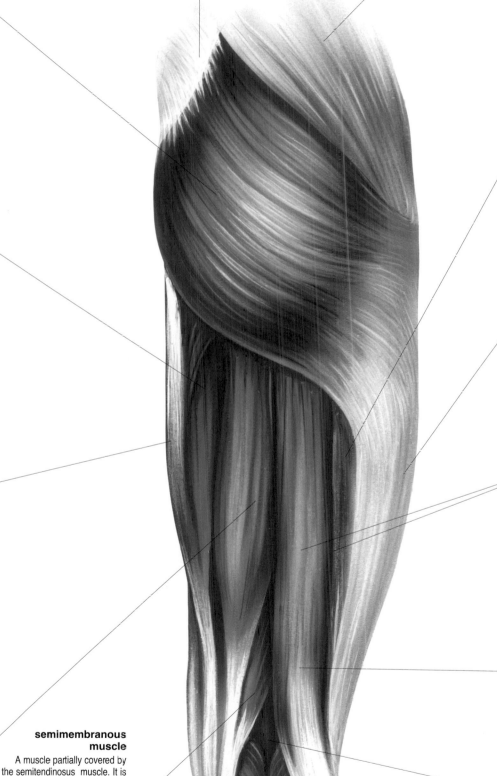

45

LEG: SUPERFICIAL MUSCLES

▼ ANTERIOR VIEW

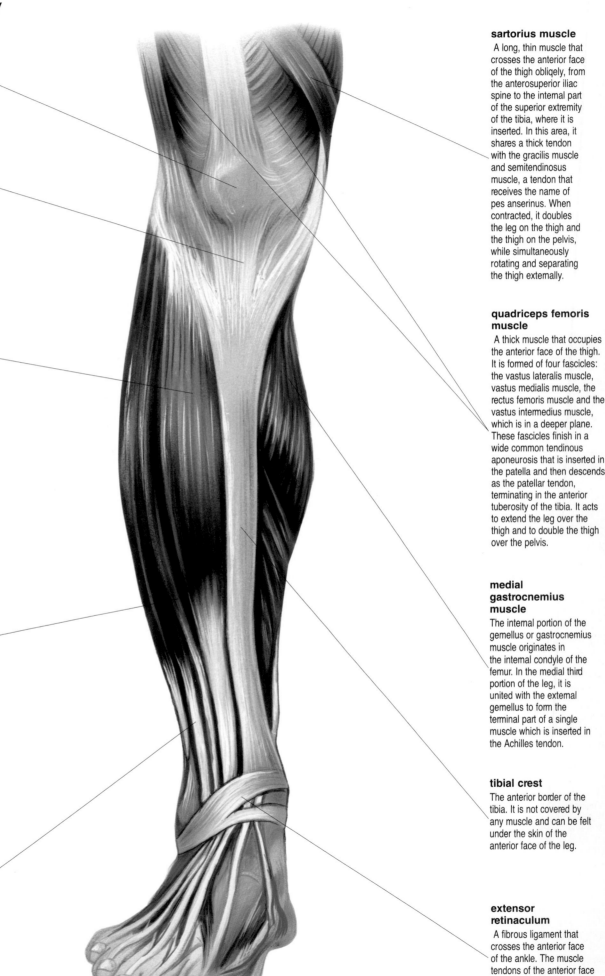

patella
A flat, round bone that occupies the anterior face of the joint of the knee.

patellar ligament
A thick ligament that extends from the vertex of the patella to the anterior tuberosity of the tibia. It is a prolongation of the tendon and is inserted inferiorly to the four portions of the quadriceps muscle.

anterior tibial muscle
A voluminous muscle that crosses the anterior face of the leg to reach the internal border of the foot. It originates in the superior extremity of the tibia. It descends and converges to become a powerful tendon that passes below the extensor retinaculum and is inserted between the first metatarsal and the first cuneiform bone. When contracted, it doubles the foot over the leg, carries the foot towards the median line and rotates the foot inwards.

extensor digitorum longus muscle
A flat muscle which shares a common origin with the anterior tibial muscle and follows a parallel trajectory until reaching the back of the foot, when it divides into four tendons that go to each of the four last toes, where they are inserted in the second and third phalanges. It serves to extend the last four toes over the dorsal face of the foot, while simultaneously doubling the foot over the leg and turning it outwards.

extensor hallucis longus muscle
A muscle partially covered by the tibial anterior and common extensor of the fingers muscles. It originates in the fibula and the interosseal ligament and, converted into a tendon, descends to the flexor retinaculum of the tarsus, which it passes under. It follows the internal border of the dorsal part of the foot and is inserted in the first and second phalanges of the big toe. It is an extensor muscle of the big toe over the foot, flexing or doubling the foot over the leg and turning the foot inwards.

sartorius muscle
A long, thin muscle that crosses the anterior face of the thigh obliqely, from the anterosuperior iliac spine to the internal part of the superior extremity of the tibia, where it is inserted. In this area, it shares a thick tendon with the gracilis muscle and semitendinosus muscle, a tendon that receives the name of pes anserinus. When contracted, it doubles the leg on the thigh and the thigh on the pelvis, while simultaneously rotating and separating the thigh externally.

quadriceps femoris muscle
A thick muscle that occupies the anterior face of the thigh. It is formed of four fascicles: the vastus lateralis muscle, vastus medialis muscle, the rectus femoris muscle and the vastus intermedius muscle, which is in a deeper plane. These fascicles finish in a wide common tendinous aponeurosis that is inserted in the patella and then descends as the patellar tendon, terminating in the anterior tuberosity of the tibia. It acts to extend the leg over the thigh and to double the thigh over the pelvis.

medial gastrocnemius muscle
The internal portion of the gemellus or gastrocnemius muscle originates in the internal condyle of the femur. In the medial third portion of the leg, it is united with the external gemellus to form the terminal part of a single muscle which is inserted in the Achilles tendon.

tibial crest
The anterior border of the tibia. It is not covered by any muscle and can be felt under the skin of the anterior face of the leg.

extensor retinaculum
A fibrous ligament that crosses the anterior face of the ankle. The muscle tendons of the anterior face of the leg pass under this ligament to the dorsal part of the foot.

LEG: SUPERFICIAL MUSCLES

▼ POSTERIOR VIEW

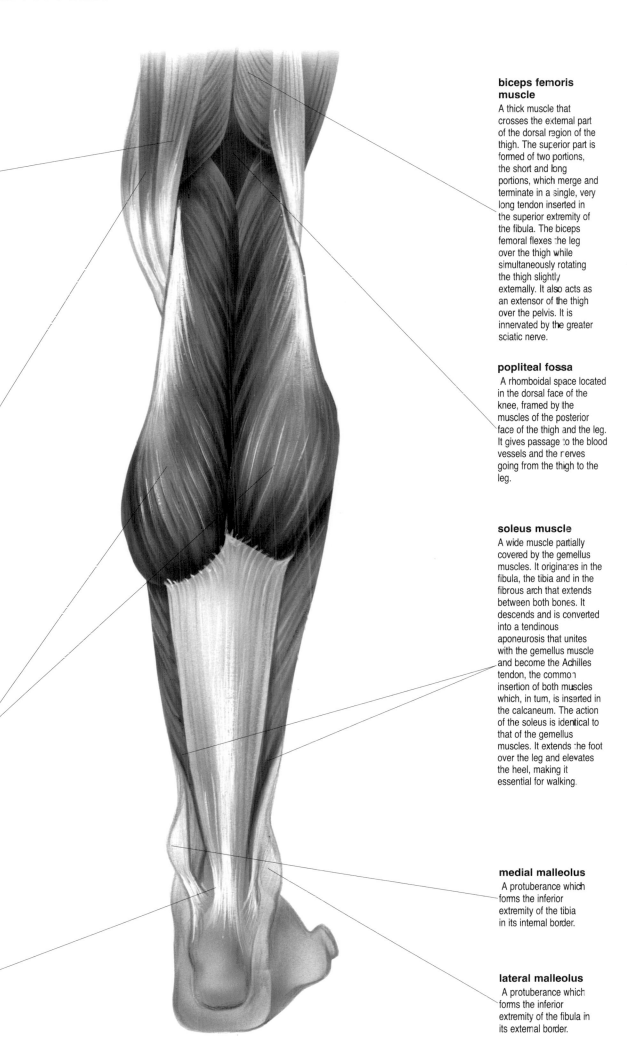

semimembranous muscle

A muscle partially covered by the semitendinosus muscle. It is called the semimabranous muscle because its superior third is composed of a wide, tendinous membrane. It originates in the ischium and is inserted in the superior extremity of the tibia. Its action is similar to the semitendinosus muscle, doubling the leg on the thigh, rotating the leg inwards and extending the thigh over the pelvis.

semitendinosus muscle

A muscle with inferior and superior parts separated by a tendon. It originates in the ischium and descends the posterior face of the thigh. After passing the internal border of the knee, it becomes anterior and is converted into the tendon known as the pes anserinus, which serves as the common insertion in the superior extremity of the tibia for two other muscles: the sartorius and the gracilis muscles. When contracted, it doubles the leg over the thigh and rotates the thigh internally. It also acts as an extensor of the thigh over the pelvis.

gastrocnemius muscle

A voluminous muscle that occupies the superficial plane of the posterior face of the leg. Superiorly it is formed by two portions, the lateral and medial heads of the gastrocnemius, which originate in the external and internal femoral condyles. At the height of the median third of the leg, they unite to form a single muscle which terminates in a tendinous aponeurosis which is connected to the tendon of the soleus muscle to form the Achilles tendon which is inserted in the calcaneum. When contracted, these muscles extend the foot over the leg. When the heel is on the ground, they lift it while simultaneously doubling the leg over the thigh. All these actions make them essential for walking.

Achilles tendon

A tendon which serves as the common insertion of the gastrocnemius and soleus muscles. It is inserted in the posterior tuberosity of the calcaneum in the posterior face of the ankle. The Achilles is a powerful tendon that can be seen under the skin of the posterior face of the ankle.

biceps femoris muscle

A thick muscle that crosses the external part of the dorsal region of the thigh. The superior part is formed of two portions, the short and long portions, which merge and terminate in a single, very long tendon inserted in the superior extremity of the fibula. The biceps femoral flexes the leg over the thigh while simultaneously rotating the thigh slightly externally. It also acts as an extensor of the thigh over the pelvis. It is innervated by the greater sciatic nerve.

popliteal fossa

A rhomboidal space located in the dorsal face of the knee, framed by the muscles of the posterior face of the thigh and the leg. It gives passage to the blood vessels and the nerves going from the thigh to the leg.

soleus muscle

A wide muscle partially covered by the gemellus muscles. It originates in the fibula, the tibia and in the fibrous arch that extends between both bones. It descends and is converted into a tendinous aponeurosis that unites with the gemellus muscle and become the Achilles tendon, the common insertion of both muscles which, in turn, is inserted in the calcaneum. The action of the soleus is identical to that of the gemellus muscles. It extends the foot over the leg and elevates the heel, making it essential for walking.

medial malleolus

A protuberance which forms the inferior extremity of the tibia in its internal border.

lateral malleolus

A protuberance which forms the inferior extremity of the fibula in its external border.

47

LEG: SUPERFICIAL MUSCLES

▼ EXTERNAL VIEW

iliotibial tract
A fibrous membrane that crosses the thigh laterally and superficially and is formed by the prolongation of the membrane covering the thigh, the fascia lata.

biceps femoris muscle
A thick muscle that crosses the external part of the dorsal region of the thigh. The superior part is formed of two portions, the short and long portions, which merge and terminate in a single, very long tendon inserted in the superior extremity of the fibula. The biceps femoral flexes the leg over the thigh, while simultaneously rotating the thigh slightly externally. It also acts as an extensor of the thigh over the pelvis.

short portion of the biceps femoral
The portion that forms the superior part of the biceps femoris muscle. It originates in the internal border of the femur in the linea aspera.

gastrocnemius
A voluminous muscle that occupies the superficial plane of the posterior face of the leg. Superiorly formed by two portions, the external gemellus and the internal gemellus, which originate in the external and internal femoral condyles. At the height of the median third of the leg, they unite to form a single muscle which terminates in a tendinous aponeurosis which joins with the tendon of the soleus muscle to form the Achilles tendon which is inserted in the calcaneum. When contracted, these muscles extend the foot over the leg. When the heel is on the ground, they lift it while simultaneously doubling the leg over the thigh. All these actions make them essential for walking.

soleus muscle
A wide muscle partially covered by the gemellus muscles which owes its name to its shape. It originates in the head, posterior face and external border of the fibula, the oblique line of the posterior face of the tibia and in the fibrous arch that extends between both bones. It descends and is converted into a tendinous aponeurosis that unites with the gemellus muscle and become the Achilles tendon, the common insertion of both muscles which, in turn, is inserted in the calcaneum. The action of the soleus is identical to that of the gemellus muscles. It extends the foot over the leg and elevates the heel, making it essential for walking.

Achilles tendon
The common insertion of the gastrocnemius and soleus muscles. It is located in the tuberosity of the calcaneum in the posterior face of the ankle.

lateral malleolus
A protuberance which forms the inferior extremity of the fibula in its external border.

fibular retinaculum
A long ligament that extends from the lateral malleolus of the fibula to the external face of the calcaneum. The tendons of the long and short peroneal muscles pass under this ligament.

quadriceps femoris muscle
A thick muscle that occupies the anterior face of the thigh. It is formed of four fascicles: the vastus lateralis muscle, vastus medialis muscle, the rectus femoris muscle and the vastus intermedius muscle, which is in a deeper plane. These fascicles finish in a wide, common tendinous aponeurosis inserted in the patella which descends as the patellar tendon, terminating in the anterior tuberosity of the tibia. Its main action extends the leg over the thigh and acts to double the thigh over the pelvis.

peroneus longus muscle
A thin muscle that occupies the external border of the leg. It originates in the fibula and the tibia and, in its inferior part, becomes a tendon that passes the lateral malleolus posteriorly to reach the sole of the foot which it crosses obliquely to be inserted in the first metatarsal. Its action extends the foot over the leg and rotates the foot outwards.

extensor digitorum longus muscle
A flat muscle which shares a common origin with the anterior tibial muscle and follows a parallel trajectory until reaching the back of the foot, where it divides into four tendons that go to each of the last four toes, where they are inserted in the second and third phalanges. It extends the last four toes over the dorsal face of the foot, while simultaneously doubling the foot over the leg and turning it outwards.

peroneus brevis muscle
A muscle located below the long peroneal muscle. It extends from the external face of the fibula to the lateral malleolus of the ankle, which it crosses posteriorly as a tendon and reaches the external area of the foot to be inserted in the fifth metatarsal. It is a separating or abductor muscle of the foot, while simultaneously rotating the foot externally.

anterior tibial muscle
A voluminous muscle that crosses the anterior face of the leg to reach the internal border of the foot. It originates in the superior extremity of the tibia. It descends and converges to become a powerful tendon that passes below the extensor retinaculum and is inserted in the first metatarsal and the first cuneiform bone. When contracted, it doubles the foot over the leg, carries the foot towards the median line and rotates the foot inwards.

peroneus tertius muscle
A small, flat muscle that originates in the inferior half of the fibula. It descends to form a tendon which passes below the flexor retinaculum of the tarsus, crossing the external border of the foot to be inserted in the fifth metatarsal. It is a flexor muscle of the foot, simultaneously rotating and separating the foot externally.

extensor retinaculum
A fibrous ligament that crosses the anterior face of the ankle. Its internal part is formed of an inferior branch and a superior branch which further divides into deep and superficial branches. The tendons of the muscles of the anterior face of the leg pass under this ligament to the dorsal part of the foot and are stabilized by it.

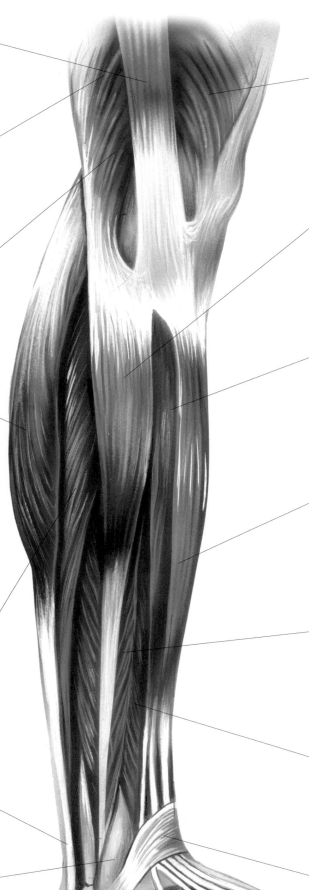

LEG: SUPERFICIAL MUSCLES

▼ INTERNAL VIEW

semitendinosus muscle

A muscle with inferior and superior parts. It originates in the ischium and descends the posterior face of the thigh. After passing the internal border of the knee, it becomes anterior and is converted into the tendon known as the pes anserinus, which serves as the common insertion in the superior extremity of the tibia for two other muscles: the sartorius and the gracilis muscles. When contracted, it doubles the leg over the thigh and rotates the thigh internally. It also acts as an extensor of the thigh over the pelvis.

patella

A flat, rounded bone that occupies the anterior face of the knee joint.

pes anserinus

A thick tendon inserted in the internal part of the superior extremity of the tibia and originates in the union of the tendons of three muscles: the sartorius, semitendinosus and the gracilis muscles. Its name means "goose's foot."

anterior tibial muscle

A voluminous muscle that crosses the anterior face of the leg to reach the internal border of the foot. It originates in the superior extremity of the tibia. It descends and converges to become a powerful tendon that passes below the extensor retinaculum and is inserted between the first metatarsal and the first cuneiform bone. When contracted, it doubles the foot over the leg, carries the foot towards the median line and rotates the foot inwards.

tibial crest

The anterior border of the tibia. It is not covered by any muscle, but is located immediately under the skin of the anterior face of the leg.

extensor retinaculum

A fibrous ligament that crosses the anterior face of the ankle. Its internal part is formed of an inferior branch and a superior branch which divide further into deep and superficial branches. The tendons muscle of the anterior face of the leg pass under this ligament to the dorsal part of the foot and are stabilized by it.

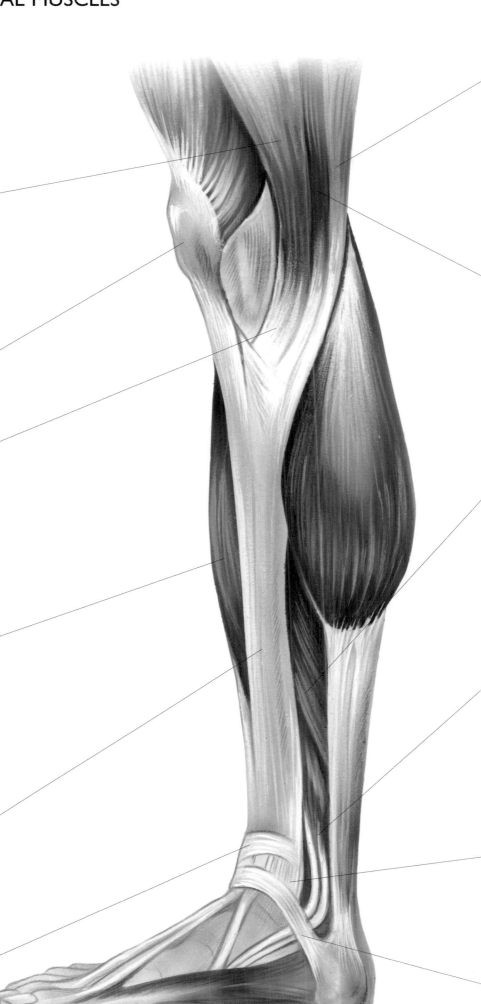

semimembranous muscle

A muscle partially covered by the semitendinosus muscle. It is called the semimabranous muscle because its superior third is composed of a wide tendinous membrane. It originates in the ischium and is inserted in the superior extremity of the tibia. It acts similarly to the semitendinosus muscle, doubling the leg over the thigh, rotating the thigh inwards and extending it over the pelvis.

gracilis muscle

A long, thin muscle that originates in the pubic bone and follows the internal border of the thigh to reach the superior part of the tibia, where it is inserted by a tendon common to two other muscles, the semitendinosus and the sartorius. The tendon is known as the pes anserinus. The gracilis muscle is a flexor of the leg over the thigh and an approximator of the thigh.

flexor digitorum longus

A muscle that originates in the posterior face of the tibia and descends as a tendon, passing behind the medial malleolus and crossing the plant of the foot, where it branches into four portions that terminate in the distal phalanges of the four last toes, which flex when contracted.

flexor hallucis longus

A muscle that follows a path parallel to the common flexor of the fingers, although it originates in the fibula. It passes behind the medial malleolus and reaches the sole of the foot as a tendon which is inserted in the second phalange of the big toe. Its function is to flex the big toe.

deltoid ligament

The articulation of the tibia with the tarsus is reinforced in its internal face by a powerful ligament whose superficial layer has a triangular shape which gives it its name. It extends from the medial malleolus of the tibia to the calcaneum, scaphoid and talus.

flexor retinaculum

A ligament which extends from the medial malleolus of the tibia to the internal face of the calcaneum. The tendons of the flexor muscles of the posterior face of the leg pass under the ligament and are stabilized by it.

49

FOOT: SUPERFICIAL MUSCLES

▼ DORSAL VIEW

lateral malleolus
A protuberance formed by the inferior extremity of the fibula in its external border.

extensor digitorum brevis
A flat muscle that originates in the calcaneous bone and crosses the dorsal area of the foot obliquely. It divides into four tendons which go to the first, second, third and fourth toes. In the first toe, the tendon is inserted in the first phalange, whereas in the other three, it is united to the tendons of the common extensor muscle of the toes. Its action extends the four first toes over the back of the foot, collaborating with the extensor digitorum longus muscle.

tendon of the anterior peroneal muscle
A tendon whose muscular mass is located in the anterior face of the leg. It crosses the external border of the foot and is inserted in the fifth metatarsal bone. It flexes the foot, simultaneously rotating it as well.

tendons of the extensor digitorum longus muscle
The four tendons of this muscle, located in the anterior face of the leg, reach the dorsal part of the foot and go to the second and third phalanges of each of the four last toes. They act to extend the last four over the dorsal part of the foot, while simultaneously doubling the foot over the leg and moving it outwards.

tendon of the anterior tibial muscle
A tendon that comes from the anterior face of the leg, passes below the extensor retinaculum and is inserted in the first metatarsal and the first cuneiform bone. When contracted, it doubles the foot on the leg, carries the foot towards the median line in adduction or approximation and rotates the foot inwards.

medial malleolus
A protuberance formed by the inferior extremity of the tibia in its internal border.

extensor retinaculum
A fibrous ligament that crosses the anterior face of the ankle. It consists of inferior and superior parts. The tendons of the muscles of the anterior face of the leg pass under this ligament to the dorsal part of the foot.

dorsal interosseal muscles
Four muscles in the deep plane of the plantar region, which can be visualized from the dorsal face. They are located in the intermetatarsal spaces and go to the first phalanges of the second, third and fourth toes. They flex the first phalanges and extend the other two phalanges of these toes.

tendon of the extensor hallucis longus muscle
A tendon whose muscular mass is in the anterior face of the leg. It crosses the internal border of the dorsal part of the foot and is inserted in the first and second phalanges of the big toe. It is an extensor muscle of the big toe over the foot and also flexes or doubles the foot over the leg, turning the foot inwards.

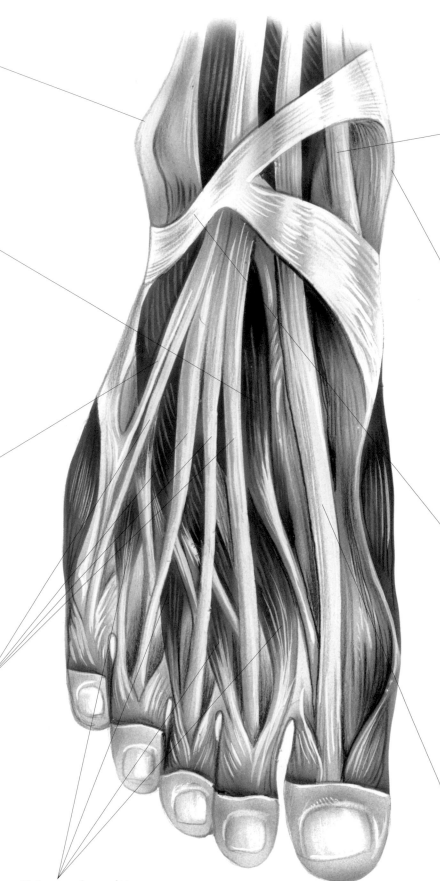

FOOT: SUPERFICIAL MUSCLES

▼ PLANTAR VIEW

flexor digitorum brevis muscle

A muscle located in the central area of the plant of the foot. It originates in the calcaneum and divides into four fleshy fascicles which become four tendons that are inserted in the second phalanges of the four last toes by thin extensions. It serves to flex the first and second phalanges of the last four toes over the sole of the foot.

abductor hallucis

A muscle occupying the internal border of the sole of the foot and extending from the calcaneum to the base of the first phalange of the big toe, where it is inserted by a tendon which is joined with the tendon of the flexor hallucis longus muscle. It flexes the big toe over the sole of the foot while simultaneously separating the big and second toes.

tendon of the flexor hallucis longus

A tendon that comes from the deep plane of the posterior face of the leg. It originates in the fibula, passes behind the medial malleolus and below the calcaneum, and is inserted in the second phalange of the big toe. It extends the foot and flexes the phalanges of the big toe.

plantar aponeurosis

A thick, triangular membrane that covers the plantar musculature immediately below the skin, from the posterior part of the calcaneum to the base of the five toes. It emits fibrous prolongations which act as sheaths for the tendons of the flexor muscles.

flexor hallucis brevis muscle

A muscle that originates in the tarsus and divides into two fascicles: one is united to the tendon of the abductor hallucis and the other to the tendon of the abductor hallucis muscle, located in the deep plane of the sole of the foot. It serves to flex the big toe over the sole of the foot.

abductor digiti minimi

A muscle located in the external border of the sole of the foot which extends from the calcaneum to the first phalange of the small toe and the fifth metatarsal. When contracted, it separates the fifth toe outwards, in a movement of abduction.

flexor digiti minimi brevis muscle

A muscle that follows a path parallel to that of the separator of the fifth toe. It extends from the base of the fifth metatarsal to the first phalange of the small toe. It collaborates with the flexor digitorum brevis muscle in the flexion of the fifth toe.

lumbrical muscles

Small, cylindrical muscles that terminate in the first phalanges of each toe. They serve to double the first phalange of the four last toes and to extend the other two phalanges.

SKELETON

▼ ANTERIOR GENERAL VIEW

mandible
A horseshoe-shaped bone, also called the jawbone, located in the inferior part of the face. Its articulation with the cranial bones allows the action of chewing.

cranium
Set of 8 flat bones (1 frontal, 2 parietal, 2 temporal, 1 occipital, 1 sphenoid and 1 ethmoid bone), whose articulation comprises the cranial cavity, which houses the brain.

skeleton of the face
Set of bones that constitute the skeleton of the face (1 vomer, 2 maxillae, 2 nasal, 2 palatine, 2 zygomatic, 2 lacrimal and 2 inferior nasal conchae).

ribs
Flat, curved bones that surround the thoracic cavity laterally from the edges of the sternum to the dorsal vertebral column, forming the rib cage. Eight ribs are fixed and the other four are joined by ligaments or have one unattached end.

1st rib
2nd rib
3rd rib
4th rib
5th rib
6th rib
7th rib
8th rib
9th rib
10th rib
11th rib
12th rib

clavicles
Two long, flat bones, located between the sternum and the scapula that serve as the fixation point of the upper limbs with the thorax.

sternum
A flat bone located in the central anterior area of the thoracic cavity, in which the ribs surrounding the cavity are inserted laterally.

ilium
A shovel-shaped part of the hip bone which articulates posteriorly with the sacrum forming the lateral wall of the pelvic cavity.

pubis
Part of the hip bone that closes the anterior part of the pelvic cavity forming a midline, cartilaginous joint called the pubic symphysis.

ischium
Part of the hip bone that serves as the lateral anterior union between the pubis and the ilium. The ischium and the ilium form an articulation which is the insertion point of the femur.

hip bone
The bone that composes the skeleton of the pelvis. It consists of 3 parts: the ilium, ischium and pubis.

femur
A long, thick bone that forms the skeleton of the thigh. In its superior part, it angles inwards to unite with the pelvis in the hip joint.

patella
A flat, triangular bone located in front of the knee joint which serves as the insertion point for many muscles of the thigh and the leg.

tarsus
A set of seven bones arranged in two rows (calcaneus, talus, cuboid, scaphoid and three cuneiform bones) that constitute the skeleton of the heel of the foot.

metatarsus
A set of five long bones that comprise the plantar vault and extend from the second row of tarsal bones to the phalanges of the fingers.

tibia
A long bone that forms the internal part of the skeleton of the leg. In its superior part, it joins the femur to form the knee joint and the inferior part unites with the tarsus and the fibula to form the ankle joint.

fibula
A long, thin bone that forms the external part of the skeleton of the leg. It articulates above and below with the tibia.

phalanges
A set of bones that form the bony skeleton of the toes. Each toe has three phalanges, except the first, which only has two.

52

SKELETON

▼ POSTERIOR GENERAL VIEW

scapula
or **shoulder bone**
A flat bone located in the
posterior face of the thorax. It
acts as the union between the
thorax and the upper limbs.

humerus
A long bone that forms the
skeleton of the arm. The
upper end articulates with the
scapula and the distal end
with the ulna and radius.

radius
The long external bone of
the forearm. Its upper
articulation is with the
humerus and the ulna in
the elbow joint and its
distal articulation with the
ulna and the carpal
bones in the wrist joint.

ulna
A long bone that forms
the internal skeleton of
the forearm. It plays an
important role in the
rotatory movements of the
forearm and hand.

carpus
A set of eight small bones
(hamate, navicular,
trapezium, pisiform,
trapezoid, lunate, triquetrum,
and capitate bones),
distributed in two rows of
four cube-shaped bones.
They articulate with the ulna
and radius in the wrist joint,
and with the metacarpal
bones of the hand.

metacarpus
A set of five long bones that
extend radially across the
hand from the carpus to the
phalanges of the fingers.

**phalanges
of the fingers**
A set of bones that form the
skeleton of the fingers. Each
of the fingers contains three
phalanges, except for the
thumb, which has only two.

vertebral column
A set of 24 bones called
vertebrae that articulate with
one another. It is divided
into 3 parts: superior or
cervical (7 vertebrae),
median or dorsal (12
vertebrae) and inferior
or lumbar (5 vertebrae).

sacrum
A triangular bone that forms the
base of the vertebral column.
Its superior articulation is with
the vertebral column and the
lateral articulation with the iliac
bone and it thus acts as an
important point of articulation,
allowing the inclination
and extension of the thorax
forwards and backwards.

coccyx
A small, terminal appendix
located at the distal end of
the sacrum and consisting of
a series of very small, almost
atrophic vertebral vestiges.

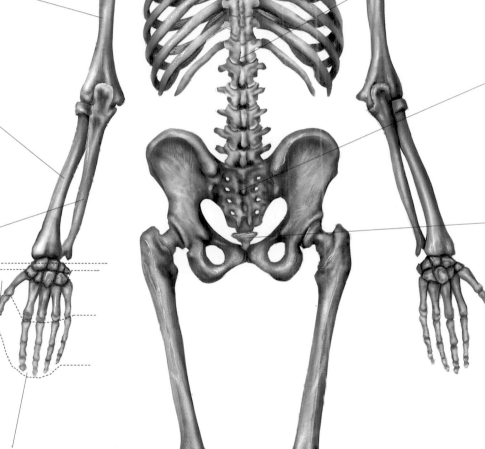

SKELETON

▶ LATERAL GENERAL VIEW

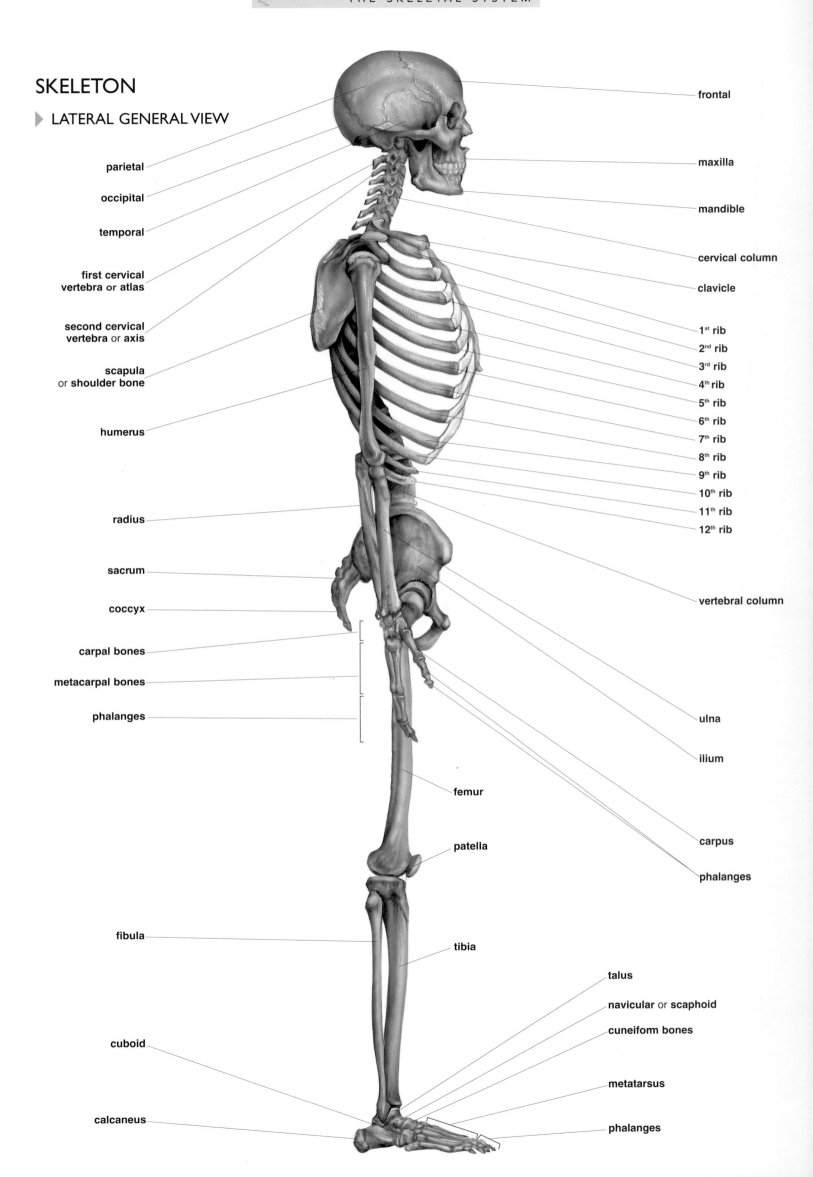

parietal

occipital

temporal

first cervical
vertebra or atlas

second cervical
vertebra or axis

scapula
or shoulder bone

humerus

radius

sacrum

coccyx

carpal bones

metacarpal bones

phalanges

fibula

cuboid

calcaneus

frontal

maxilla

mandible

cervical column

clavicle

1st rib

2nd rib

3rd rib

4th rib

5th rib

6th rib

7th rib

8th rib

9th rib

10th rib

11th rib

12th rib

vertebral column

ulna

ilium

carpus

phalanges

femur

patella

tibia

talus

navicular or scaphoid

cuneiform bones

metatarsus

phalanges

54

STRUCTURE OF A LONG BONE

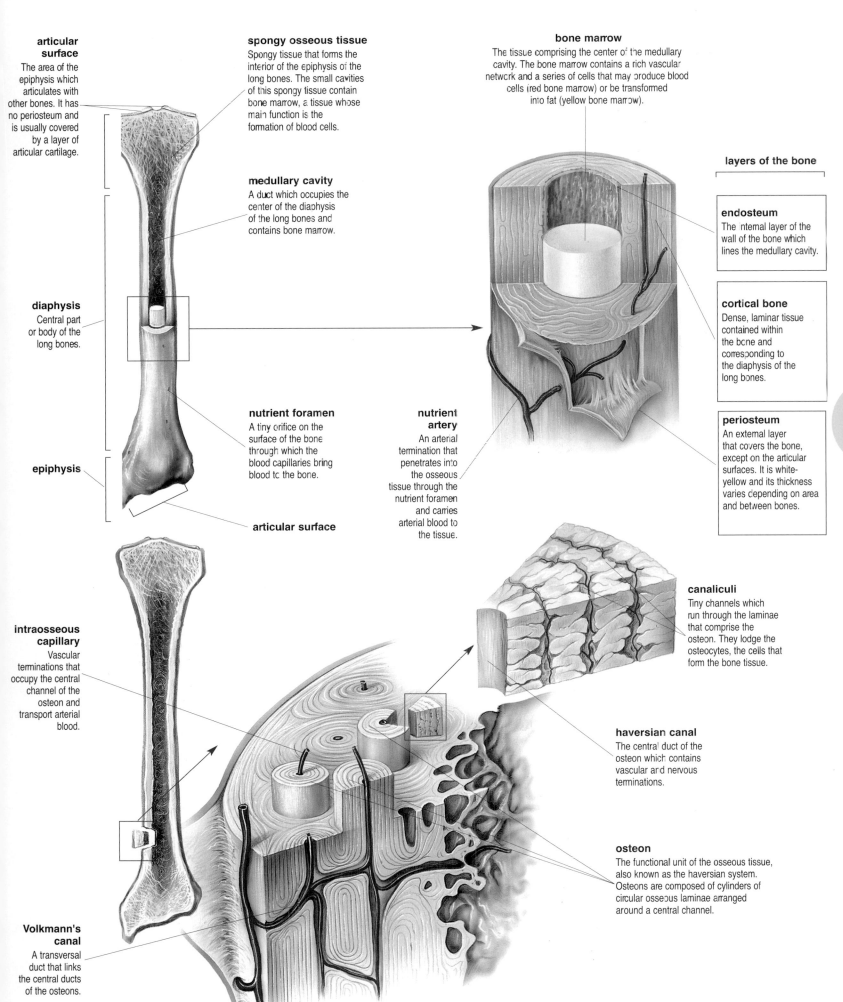

articular surface
The area of the epiphysis which articulates with other bones. It has no periosteum and is usually covered by a layer of articular cartilage.

spongy osseous tissue
Spongy tissue that forms the interior of the epiphysis of the long bones. The small cavities of this spongy tissue contain bone marrow, a tissue whose main function is the formation of blood cells.

bone marrow
The tissue comprising the center of the medullary cavity. The bone marrow contains a rich vascular network and a series of cells that may produce blood cells (red bone marrow) or be transformed into fat (yellow bone marrow).

layers of the bone

endosteum
The internal layer of the wall of the bone which lines the medullary cavity.

medullary cavity
A duct which occupies the center of the diaphysis of the long bones and contains bone marrow.

diaphysis
Central part or body of the long bones.

cortical bone
Dense, laminar tissue contained within the bone and corresponding to the diaphysis of the long bones.

nutrient foramen
A tiny orifice on the surface of the bone through which the blood capillaries bring blood to the bone.

nutrient artery
An arterial termination that penetrates into the osseous tissue through the nutrient foramen and carries arterial blood to the tissue.

periosteum
An external layer that covers the bone, except on the articular surfaces. It is white-yellow and its thickness varies depending on area and between bones.

epiphysis

articular surface

55

intraosseous capillary
Vascular terminations that occupy the central channel of the osteon and transport arterial blood.

canaliculi
Tiny channels which run through the laminae that comprise the osteon. They lodge the osteocytes, the cells that form the bone tissue.

haversian canal
The central duct of the osteon which contains vascular and nervous terminations.

osteon
The functional unit of the osseous tissue, also known as the haversian system. Osteons are composed of cylinders of circular osseous laminae arranged around a central channel.

Volkmann's canal
A transversal duct that links the central ducts of the osteons.

SKULL

▼ FRONTAL VIEW

frontal bone
A large bone that forms the anterior part of the cranium and lodges, in its posterior concave face, the frontal lobes of the brain.

sphenoid bone
One of the bones that form the cranial base. In its anterior part, it forms part of the wall of the orbit.

orbits
Two cavities located in the superior part of the facial skeleton which lodge the eyeballs. They are limited by the frontal bone, the sphenoid bone, the zygomatic bones, the maxillae and the lacrimal bones.

orbital fissure
An irregular cleft between the greater and lesser wings of the sphenoid bone, through which the nerves and blood vessels of the eye pass.

superciliary arch
The arches which form the superior borders of the orbits and are the bony support of the eyebrows.

zygomatic bone
One of the bones that forms the facial skeleton. It constitutes the bony support of the cheeks. Its superior edges form the floor and the external face of the orbit.

nasal septum
The osteocartilaginous plate that divides the nasal fossas. It is formed in its posterior part by the vomer bone.

maxillae
Two symmetrical bones that delimit the cavity of the nasal fossas and are united in their inferior part. They also form part of the floor of the orbit and are the center of the facial skeleton. In their inferior face, they lodge the upper teeth.

mandibular angle
The angle that forms at each side of the cranium where the rami and body of the mandible meet.

teeth
Extremely hard structures arranged in two rows, with the superior row being lodged in the maxilla and the inferior row in the mandible.

mandible
As the only movable bone of the head, the mandible serves as the inferior part of the face. It is articulated with the temporal bone by a joint that allows the action of chewing. The mandible has two sections: the perpendicular rami and the horizontal body.

56

SKULL

▼ LATERAL VIEW

frontosphenoidal suture
The line formed where the frontal bone meets the ala of the sphenoid bone.

external acoustic meatus
A bony canal that forms the opening in the external temporal bone which communicates the cavity of the middle ear with the exterior.

coronal suture
An articulation that unites the posterior edge of the frontal bone with the anterior edges of the parietal bones.

parietal bone
Two symmetrical bones located on both sides of the cranium and united by the sagittal suture.

temporoparietal suture
A fixed articulation which unites the squama of the temporal bone with the parietal bone.

temporal bone
A complex bone that contributes structurally to the cranial vault. It contains the hearing organs and is composed of five parts: the tympanic, petrous and mastoid parts, the styloid process and the squama.

lamboid suture
The lamboid suture articulates the posterior edge of the parietal bone with the occipital bone.

ala of the sphenoid bone
A prolongation of the sphenoid bone that forms part of the temporal fossa.

occipital bone
A single, concave bone forming the posteroinferior part of the cranium. It houses the occipital lobes of the brain and the cerebellum.

nasal bones
Two flat bones, articulated by their internal faces, which form the bony support of the base of the nose.

anterior nasal spine
The anterior prolongation of the maxillae which forms a tip and is located in the superior part of the intermaxillary suture.

mastoid process
The inferior protuberance of the temporal bone, which contains some irregular cavities called mastoid cells. The mastoid process plays an important role in hearing and serves as the insertion point for many neck muscles.

glenoid cavity
The deep fossa located at the base of the temporal bone which articulates with the mandibular condyle to form the temporomandibular joint.

condylar process of the mandible
A bony elevation of the ramus of the mandible which articulates with the glenoid cavity of the temporal bone, forming the temporomandibular joint.

zygomatic arch
A bony arch which originates in the zygomatic bone and extends laterally backwards, where it is united with the temporal bone.

coronoid process
A bony elevation located in the anterior part of the superior border of the rami of the mandible where the temporal muscle is inserted, allowing the mandible to move.

sigmoid notch
A wide notch extending from the coronoid process to the mandibular condyle which allows the passage of vessels and nerves.

temporal fossa
A bony depression located in the lateral face of the cranium and formed by the alas of the sphenoid bone and the concha of the temporal bone. A large part of the temporal bone is inserted in the fossa.

styloid process
A long, thin prolongation of the inferior face of the temporal bone which serves as the insertion point for ligaments and muscles.

57

CRANIAL VAULT

▼ EXTERNAL VIEW

▼ INTERNAL VIEW

lamboid suture
The lamboid suture articulates the posterior edge of the parietal bone with the occipital bone.

occipital bone
A single, concave bone forming the posteroinferior part of the cranium. It lodges the occipital lobes of the brain and the cerebellum.

grooves for vascular branches
Traces formed in the bone by the different blood vessels surrounding the brain.

occipital fossa
The internal concavity of the occipital bone which houses the superior part of the occipital lobe of the brain.

parietal fossas
The internal concavities of the parietal bones which house the parietal lobes of the brain.

coronal suture
An articulation that unites the posterior edge of the frontal bone with the anterior edges of the parietal bones.

frontal bone
A large bone that forms the anterior part of the cranium and lodges, in its posterior concave face, the frontal lobes of the brain.

saggital suture
A line of articulation that crosses the cranial vault from the frontal bone to the occipital bone, uniting the parietal bones.

parietal bone
Two symmetrical bones located on both sides of the cranium and united by the saggital suture.

longitudinal channel
A deep sulcus that serves as a prolongation of the frontal crest. It houses the superior longitudinal sinus, an important branch of the venous system of the superior region of the cranium.

frontal crest
A bony elevation located in the central internal face of the frontal bone. It contains the falx cerebri, a membranous partition that separates the two cerebral hemispheres.

frontal fossa
Internal concavity of the frontal bone in its superior part which houses the frontal lobes of the brain.

BASE OF THE CRANIUM

▼ EXTERNAL VIEW   ▼ INTERNAL VIEW

condyles of the occipital bone
Elevations located at both sides of the foramen magnum which articulate with the atlas.

occipital bone
A single bone forming the inferior posterior part of the cranial vault, which curves to form a concave internal surface.

temporal bone
A complex bone that contributes structurally to the cranial vault. It contains the organs of hearing and is composed of five parts: the tympanic, petrous and mastoid parts, the styloid process and the squama.

external acoustic meatus
A bony canal forming the opening in the external temporal bone which communicates the cavity of the middle ear with the exterior.

mastoid process
The inferior protuberance of the temporal bone, which contains cavities called mastoid cells. The mastoid process plays an important role in hearing.

internal auditory meatus
An orifice that opens in the internal face of the petrous bone and allows the passage of the auditory and facial nerves.

basilar part of occipital bone
An inclined channel that ends at the foramen magnum and houses the medulla oblongata.

cerebellar fossa
The depression of the interior of the occipital bone posterior to the foramen magnum which houses the cerebellum.

sella turcica
A bony structure of the sphenoid bone which partly surrounds the pituitary gland and is also known as the pituitary fossa.

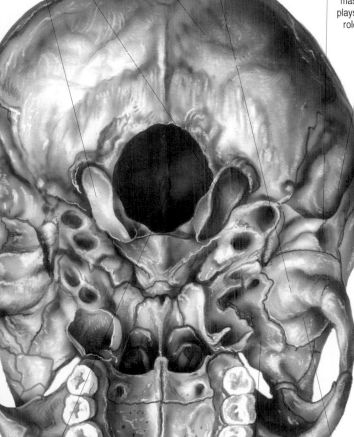

foramen magnum
An oval orifice in the occipital bone which connects the cranial cavity to the spine. The medulla oblongata merges with the spinal cord at the foramen magnum.

teeth
The 16 upper teeth are inserted in the alveolar cavities of the maxilla.

sphenoid bone
One of the bones forming the cranial base. It emits prolongations called the alas of the sphenoid which comprise part of the temporal fossa.

glenoid cavity
The deep fossa located at the base of the temporal bone which articulates with the mandibular condyle to form the temporomandibular joint.

cribiform lamina
Part of the ethmoid bone that communicates with the cranial cavity by means of tiny orifices through which branches of the olfactory nerve pass.

optic foramen
An orifice located in the internal face of the sphenoid bone which allows the ophthalmic nerve and artery to reach the orbit.

lesser alas of the sphenoid bone
Two lateral prolongations of the sphenoid bone that form the posterior wall of the orbit.

maxillae
Twin bones that delimit the nasal fossas and join to form the vault of the hard palate, or roof of the mouth.

frontal crest
A bony elevation located in the central internal face of the frontal bone. It contains the falx cerebri, a membranous partition that separates the two cerebral hemispheres in this area.

frontal bone
A large bone forming the anterior part of the cranium and lodging the frontal lobes of the brain in its posterior concave face.

VERTEBRAL COLUMN

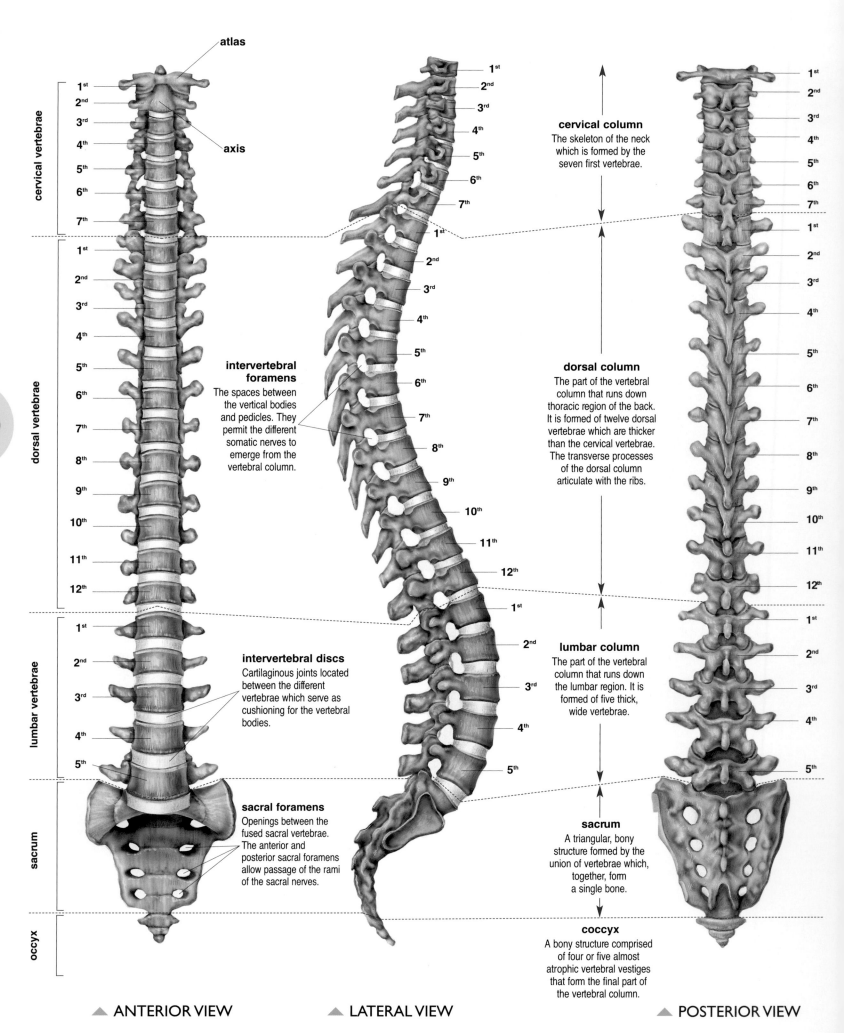

atlas

axis

cervical vertebrae

1st
2nd
3rd
4th
5th
6th
7th

dorsal vertebrae

1st
2nd
3rd
4th
5th
6th
7th
8th
9th
10th
11th
12th

lumbar vertebrae

1st
2nd
3rd
4th
5th

sacrum

occyx

60

intervertebral foramens
The spaces between the vertical bodies and pedicles. They permit the different somatic nerves to emerge from the vertebral column.

intervertebral discs
Cartilaginous joints located between the different vertebrae which serve as cushioning for the vertebral bodies.

sacral foramens
Openings between the fused sacral vertebrae. The anterior and posterior sacral foramens allow passage of the rami of the sacral nerves.

cervical column
The skeleton of the neck which is formed by the seven first vertebrae.

dorsal column
The part of the vertebral column that runs down thoracic region of the back. It is formed of twelve dorsal vertebrae which are thicker than the cervical vertebrae. The transverse processes of the dorsal column articulate with the ribs.

lumbar column
The part of the vertebral column that runs down the lumbar region. It is formed of five thick, wide vertebrae.

sacrum
A triangular, bony structure formed by the union of vertebrae which, together, form a single bone.

coccyx
A bony structure comprised of four or five almost atrophic vertebral vestiges that form the final part of the vertebral column.

Lateral view labels:
1st 2nd 3rd 4th 5th 6th 7th (cervical)
1st 2nd 3rd 4th 5th 6th 7th 8th 9th 10th 11th 12th (dorsal)
1st 2nd 3rd 4th 5th (lumbar)

Posterior view labels:
1st 2nd 3rd 4th 5th 6th 7th (cervical)
1st 2nd 3rd 4th 5th 6th 7th 8th 9th 10th 11th 12th (dorsal)
1st 2nd 3rd 4th 5th (lumbar)

▲ ANTERIOR VIEW   ▲ LATERAL VIEW   ▲ POSTERIOR VIEW

DIFFERENT TYPES OF VERTEBRAE

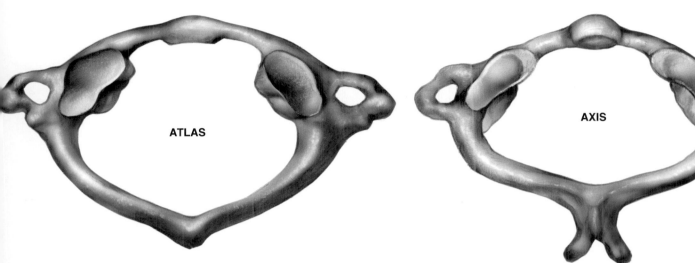

ATLAS

AXIS

The first vertebra of the cervical column. Unlike the other vertebrae, it is made up of two lateral parts united by anterior and posterior arches, presenting two glenoid cavities which articulate with the occipital bone. The small face of the anterior arch articulates with the odontoid process of the axis. The transverse foramen or holes give passage to the vertebral arteries.

The second cervical vertebra is differentiated from the other vertebrae by the fact that the vertebral body contains a prolongation known as the odontoid process, which articulates perpendicularly upwards with the atlas. The spinous process is bifurcated. Like the atlas, the axis has two lateral transverse foramen which give passage to the vertebral arteries.

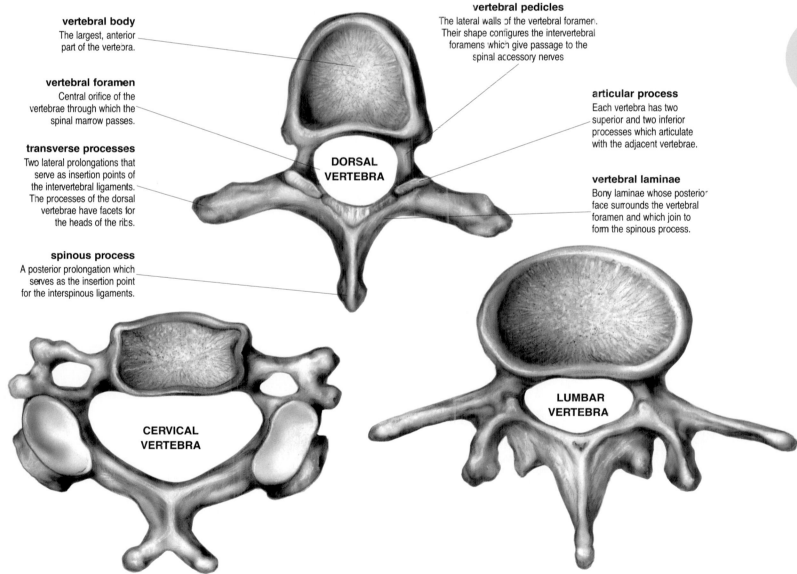

vertebral body
The largest, anterior part of the vertebra.

vertebral foramen
Central orifice of the vertebrae through which the spinal marrow passes.

transverse processes
Two lateral prolongations that serve as insertion points of the intervertebral ligaments. The processes of the dorsal vertebrae have facets for the heads of the ribs.

spinous process
A posterior prolongation which serves as the insertion point for the interspinous ligaments.

vertebral pedicles
The lateral walls of the vertebral foramen. Their shape configures the intervertebral foramens which give passage to the spinal accessory nerves

articular process
Each vertebra has two superior and two inferior processes which articulate with the adjacent vertebrae.

vertebral laminae
Bony laminae whose posterior face surrounds the vertebral foramen and which join to form the spinous process.

DORSAL VERTEBRA

CERVICAL VERTEBRA

LUMBAR VERTEBRA

The differentiating characteristics of the cervical vertebrae are: the quadrangular, vertebral body with the transversal diameter predominating; like the atlas and the axis, there are transverse holes through which the vertebral arteries pass; the spinous process is short and bifurcated and the transverse processes are implanted at the sides of the vertebral body and are short.

The characteristics of lumbar vertebrae are: a very large and high body and well-developed spinous processes which descend obliquely. The lumbar vertebrae are heavily built to withstand more weight.

THORAX

▼ ANTERIOR VIEW

▼ POSTERIOR VIEW

manubrium
The superior part of the sternum, its superior border presents a notch, the sternal notch, that can be felt under the skin. Laterally, it presents two articulate surfaces on each side, for the clavicle and the first costal cartilage.

scapula
Also known as the shoulder bone. A flat, triangular bone which joins the upper limb or extremity to the posterior part of the thorax.

clavicle
A flat, elongated bone that acts as the fixation point of the superior extremity to the sternum.

costal cartilages
Cartilaginous elements which join the ribs and the sternum.

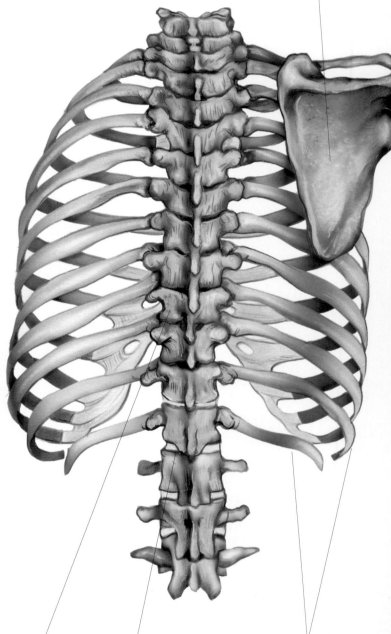

ribs
Twelve flat bones located on each side of the thorax that extend from the dorsal vertebral column to the sternum. The ribs that connect directly with the sternum are called true ribs and those which connect through a common cartilage are called false ribs.

sternum
A flat bone located in the anterior face of the thorax that serves as the anterior union of the ribs of both sides. It is divided into three parts: the manubrium, the body and the xiphoid appendix.

xiphoid appendix
Inferior end or point of the sternum which is formed of cartilaginous tissue.

body of sternum
Central segment of the sternum. It contains lateral articular faces which unite with the costal cartilages to form the sternocostal joints.

dorsal vertebral column
The bony structure formed by the twelve dorsal vertebrae. It provides posterior support to the pairs of ribs.

intervertebral discs
Cartilaginous joints located between the different vertebrae which serve as the cushioning of the vertebral bodies.

floating ribs
The last two last ribs are called floating ribs as they do not join with the sternum, leaving their anterior border free.

62

SHOULDER AND ARM

▼ ANTERIOR VIEW

▼ POSTERIOR VIEW

clavicle
A flat, elongated bone that acts as the fixation point of the superior extremity to the sternum.

greater tubercle
A large elevation located in the external zone of the neck of the humerus, which serves as an insertion point for the muscles, joining the humerus to the scapula

coracoid process
A short, bony projection of the scapula. It is tipped and serves as the insertion point for the muscles and ligaments of the shoulder and arm.

lesser tubercle
An elevation smaller than the greater tubercle located in the anterior zone of the neck of the humerus.

acromion
The projection of the scapula which constitutes the point of the shoulder. The anterior face presents a facet which articulates with the clavicle.

spine of the scapula
The posterior face of the scapula is divided by the spine of the scapula into the supraspinous and subspinous fossas.

head of the humerus
Flat, smooth surface which forms almost a third of a sphere and which articulates with the glenoid cavity of the scapula.

63

scapula
A flat, triangular bone which forms the pectoral girdle with the clavicle.

glenoid cavity
Oval articular face located at the upper lateral angle of the scapula which articulates with the head of the humerus.

anatomical neck of the humerus
A constricted area just below the head of the humerus which joins the head to the body.

humerus
A long, heavy bone that forms the skeleton of the arm. It consists of the head, which is articulated with the scapula to form the shoulder joint, the body and the lower extremity which articulates with the ulna and radius, forming the elbow joint.

medial epicondyle
A projection of the inner part of the distal extreme of the humerus, also known as the internal condyle.

olecranon fossa
A cavity located over the trochlea in its posterior face. It houses the extremity of the olecranon, or the superior part of the articular extremity of the ulna when extended.

humeral condyle
A hemispheric projection located in the external part of the lower extremity of the humerus. It articulates with the head of the radius.

coronoid fossa
A cavity located over the trochlea in its anterior face which receives the coronoid process of the ulna when the elbow is flexed.

humeral trochlea
An articular surface in the form of a pulley, located in the lower extremity of the humerus, which articulates with the sigmoid cavity of the ulna.

FOREARM

▼ ANTERIOR VIEW

greater sigmoid cavity
Hook-shaped articular cavity which articulates with the humeral trochlea.

coronoid process
The anterior end of the greater sigmoid cavity, which ends in a tip that articulates with the coronoid fossa of the humerus when the forearm is flexed.

radial tuberosity
A bony protuberance located below the neck of the radius that serves as the insertion for one of the tendons of the biceps muscle of the arm.

neck of the radius
A somewhat narrower part of the radius that joins the body of the bone to the head.

head of the ulna
A thickened, hemispheric area which forms the inferior extreme of the ulna. In its external part, it articulates with the radius and in the inferior part with the pyramidal bone of the carpus.

styloid process of the radius
A thick prolongation located in the inferior extremity of the radius and which can be felt beneath the skin. It serves as an insertion point for ligaments and muscles of the wrist and forearm.

ULNA
The long bone that occupies the inner part of the forearm and plays a fundamental role in the rotation of the forearm and hand. It consists of a central body and two extremes. The superior extreme is very voluminous and forms part of the articulation of the elbow while the inferior extreme is part of the wrist joint.

styloid process of the ulna
A cylindrical prolongation that extends vertically downwards from the inferior extremity of the ulna and serves as the insertion point for some of the ligaments of the wrist joint.

▼ POSTERIOR VIEW

olecranon
Superior extreme of the greater sigmoid cavity which finishes in a tip articulated in the olecranon fossa of the humerus.

head of radius
The superior extremity of the radius is cylindrical. The superior face is concave and is called the glenoid cavity of the radius. The internal part of the cylinder articulates with the ulna.

RADIUS
A long bone that forms the external part of the skeleton of the forearm. It consists of a central body and two extremes. The superior extreme forms part of the elbow joint and the larger, inferior extreme forms part of the wrist joint.

HAND

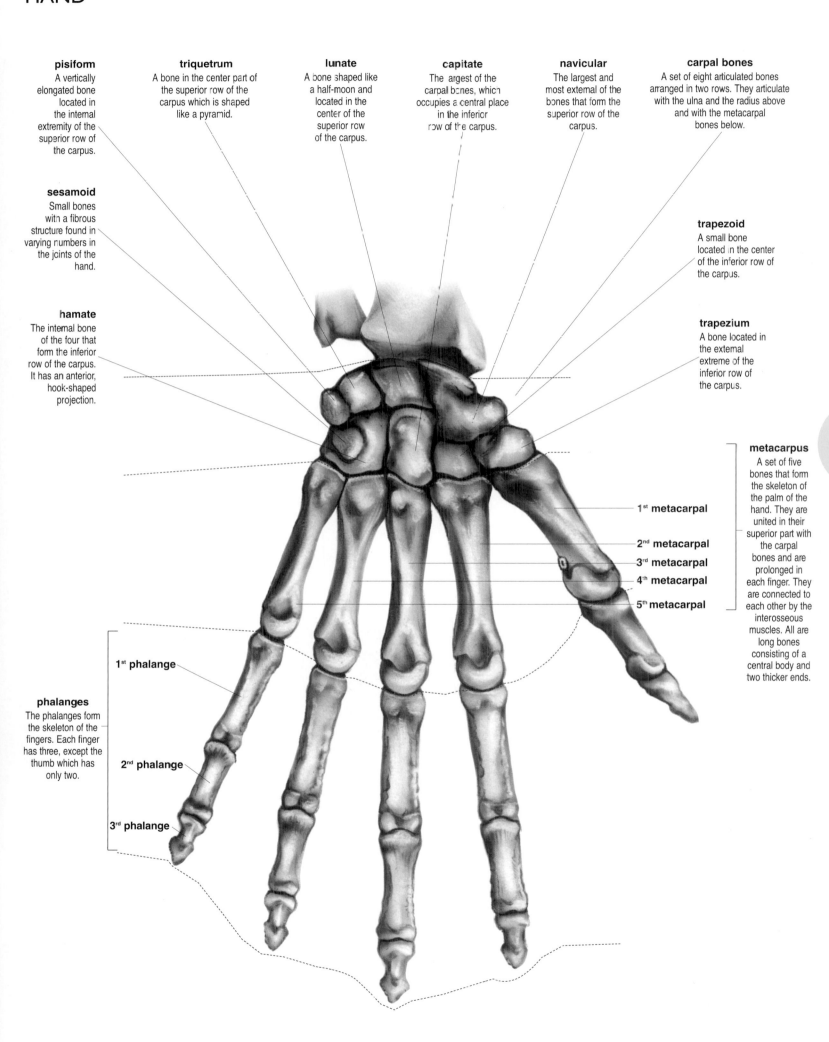

pisiform
A vertically elongated bone located in the internal extremity of the superior row of the carpus.

triquetrum
A bone in the center part of the superior row of the carpus which is shaped like a pyramid.

lunate
A bone shaped like a half-moon and located in the center of the superior row of the carpus.

capitate
The argest of the carpal bones, which occupies a central place in the inferior row of the carpus.

navicular
The largest and most external of the bones that form the superior row of the carpus.

carpal bones
A set of eight articulated bones arranged in two rows. They articulate with the ulna and the radius above and with the metacarpal bones below.

sesamoid
Small bones with a fibrous structure found in varying numbers in the joints of the hand.

trapezoid
A small bone located in the center of the inferior row of the carpus.

hamate
The internal bone of the four that form the inferior row of the carpus. It has an anterior, hook-shaped projection.

trapezium
A bone located in the external extreme of the inferior row of the carpus.

metacarpus
A set of five bones that form the skeleton of the palm of the hand. They are united in their superior part with the carpal bones and are prolonged in each finger. They are connected to each other by the interosseous muscles. All are long bones consisting of a central body and two thicker ends.

1st **metacarpal**
2nd **metacarpal**
3rd **metacarpal**
4th **metacarpal**
5th **metacarpal**

1st **phalange**

2nd **phalange**

3rd **phalange**

phalanges
The phalanges form the skeleton of the fingers. Each finger has three, except the thumb which has only two.

65

PELVIS

▼ POSTERIOR VIEW

hip bone
The three bones of the hip, the ilium, ischium and pubis form the innominate bone. The two symmetrical innominate bones form the hip bone, also called the pelvic girdle or pelvis. The bones are united posteriorly with the sacrum and anteriorly with each other by the pubic symphysis.

ilium
The external part of the hip bone, which is shaped like a shovel. It forms the lateral wall of the pelvic cavity.

ischium
The ischium forms the inferior part of the hip bone and consists of a voluminous body and an ascending branch which is united with the descending branch of the pubis.

pubis
The internal part of the hip bone. The pubis consists of a central body, a horizontal branch that articulates with the cotyloid cavity and a descending branch. The two pubic bones are connected.

iliac crest
A bony crest that forms the superior border of the ilium and extends from the sacroiliac joint to the anterior superior iliac spine.

external iliac fossa
A wide surface located in the posterior part of the ilium which serves as an insertion point for the gluteus muscles.

sciatic spine
An elevation located below the greater sciatic notch which serves as an insertion point for muscles and ligaments.

sacrum
A triangular bony structure formed by the union of five vertebrae that forms a single bone. It is located at the base of the vertebral column and articulates laterally with the hip bones.

sacroiliac joint
A joint with almost no mobility which unites the articular face of the sacrum with an ear-shaped facet of the internal part of the ilium called the auricular surface.

greater sciatic notch
A large notch located in the posterior border of the ilium through which a large number of blood vessels and nerves leave the pelvis.

coccyx
A rudimentary structure which is the vestige of the tail in human beings. It is formed of four or five atrophic fused vertebrae.

obturator foramen
Also called the lesser sciatic foramen. A large orifice located below the cotyloid cavity which is bordered by the ischium and the pubis. It is covered by a fibrous lamina called the obturator membrane.

▼ ANTERIOR VIEW

iliac crest
A bony projection located in the external border of the ilium which can be felt through the skin of the hip.

anterior inferior iliac spine
A bony projection that appears below the iliac crest. It serves as the insertion point for a muscular tendon.

interior iliac fossa
A triangular surface which serves as the insertion point for the iliac muscle and corresponds to the internal part of the ilium.

pubic symphysis
A joint which articulates the two pubic bones and closes the anterior face of the pelvic cavity.

cotyloid cavity or acetabulum
Cavity located in the center of the hip bone, forming a ball-and-socket joint with the femoral head. It is surrounded by the ilium (superior zone), the pubis (anterior inferior zone) and the ischium (posterior inferior zone).

THIGH AND KNEE

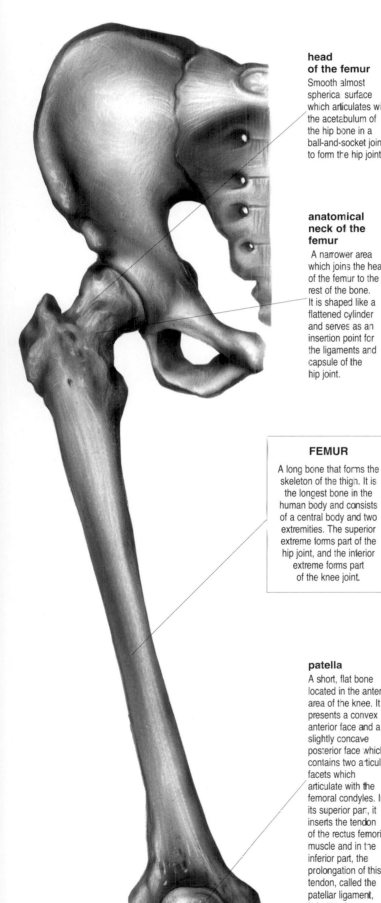

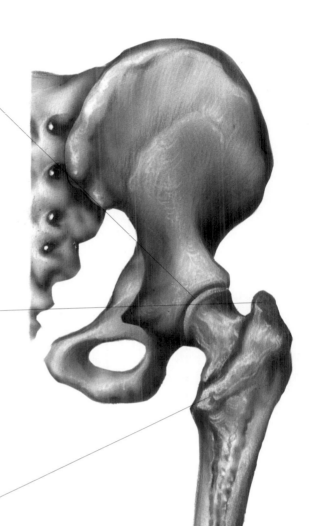

head of the femur
Smooth almost spherical surface which articulates with the acetabulum of the hip bone in a ball-and-socket joint to form the hip joint.

anatomical neck of the femur
A narrower area which joins the head of the femur to the rest of the bone. It is shaped like a flattened cylinder and serves as an insertion point for the ligaments and capsule of the hip joint.

FEMUR
A long bone that forms the skeleton of the thigh. It is the longest bone in the human body and consists of a central body and two extremities. The superior extreme forms part of the hip joint, and the inferior extreme forms part of the knee joint.

patella
A short, flat bone located in the anterior area of the knee. It presents a convex anterior face and a slightly concave posterior face which contains two articular facets which articulate with the femoral condyles. In its superior part, it inserts the tendon of the rectus femoris muscle and in the inferior part, the prolongation of this tendon, called the patellar ligament, is inserted.

hip joint
The joint formed by the head of the femur and the cotyloid cavity or acetabulum of the hip bone.

greater trochanter
A thick eminence located in the posterior external part of the base of the anatomical neck of the femur. Serves as the insertion point for several groups of muscles.

lesser trochanter
An elevation located in the posterior internal part of the anatomical neck of the femur. Serves as the insertion point for some of the muscles that unite the pelvis with the femur.

femoral trochlea
Articular surface in the shape of a pulley with a central notch that extends along the inferior extreme of the humerus and articulates with the superior part of the tibia.

medial condyle
Tuberosity located in the internal part of the lower extremity of the femur which articulates with the internal glenoid cavity of the tibia. In its lateral zone, it presents a tuberosity that serves as the place of insertion for the knee ligaments.

lateral condyle
A tuberosity located in the external part of the lower extremity of the femur which articulates with the lateral articular surface of the tibia. It presents a lateral tuberosity which inserts articular ligaments.

67

▲ ANTERIOR VIEW

▲ POSTERIOR VIEW

LEG

head of the fibula
The superior part of the fibia is larger than the rest. The internal part contains an articular facet which articulates with the upper extremity of the tibia.

medial intercondylar tuberule
An elevated projection that separates the lateral and medial articular faces of the tibia and articulates with the femoral trochlea.

articular cavities of the tibia
Two slightly concave articular faces, one lateral and the medial, located in the superior face of the tibia, which articulate the lateral and medial condyles of the femur, respectively.

anterior tibial tuberosity
A projection located in the superior part of the anterior border of the tibia in which the patellar ligament is inserted.

styloid process of the fibula
A bony projection which extends vertically upwards from the head of the fibula. It inserts a tendon of the biceps femoris muscle.

tibial crest
The anterior border of the tibia which crosses the anterior part of the body of the bone longitudinally. It has no muscular insertions and can be felt through the skin.

FIBULA
A long bone that forms the external part of the skeleton of the leg. It has a body and two extremes, the superior and inferior, that articulate with the tibia, to form the superior and inferior tibiofibular joints. The inferior extreme also articulates with the talus.

TIBIA
A long bone that comprises the internal skeleton of the leg. It has a long central body and two extremes. The large, superior extreme forms part of the knee joint and the smaller, inferior extreme forms part of the ankle joint.

lateral malleolus of the fibula
A bony projection located in the external part of the inferior extremity of the fibula, immediately below the skin. It is furrowed by a channel that gives passage to the tendons of the muscles of the fibula.

medial tibial malleolus
A thick projection located in the internal area of the tibia and forming part of the ankle joint. It serves as an insertion point for various ligaments.

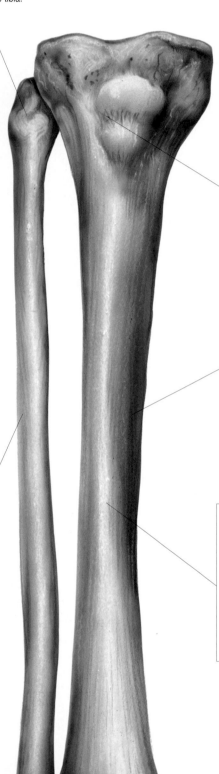

▲ ANTERIOR VIEW

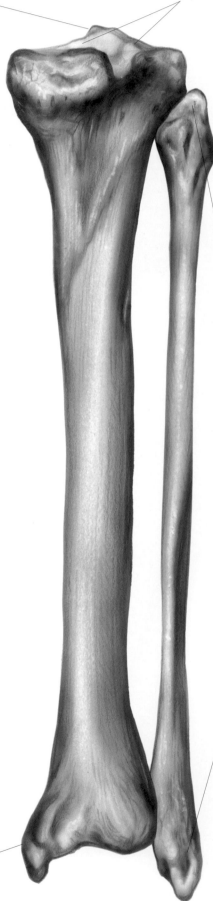

▲ POSTERIOR VIEW

FOOT

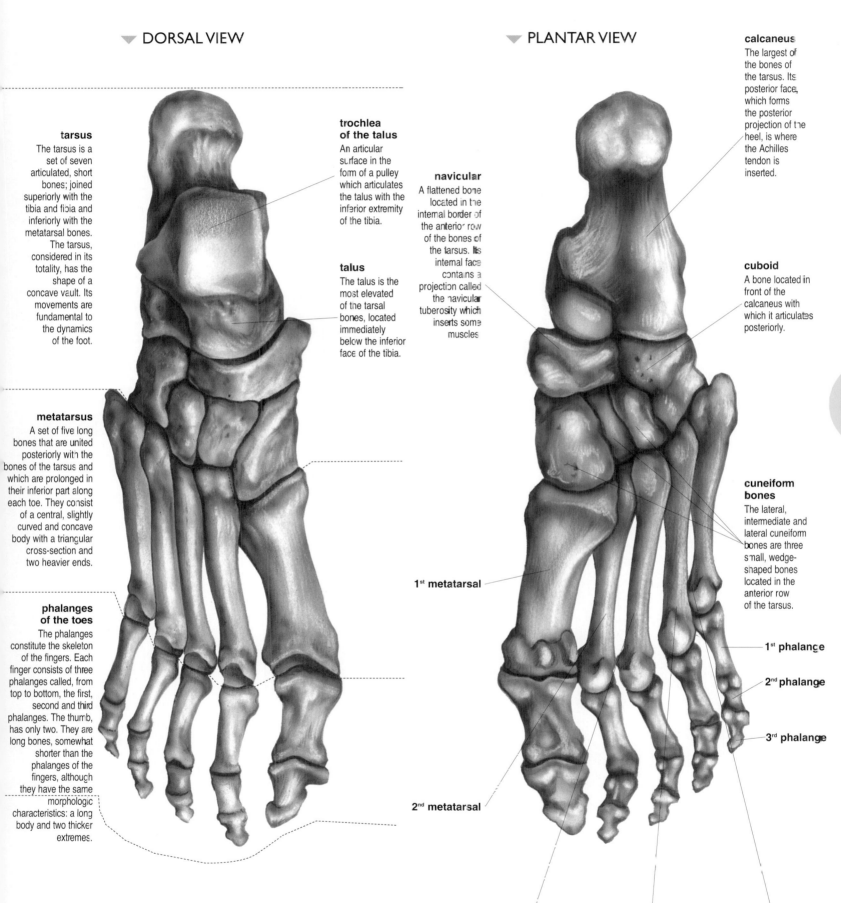

tarsus
The tarsus is a set of seven articulated, short bones; joined superiorly with the tibia and fibula and inferiorly with the metatarsal bones. The tarsus, considered in its totality, has the shape of a concave vault. Its movements are fundamental to the dynamics of the foot.

trochlea of the talus
An articular surface in the form of a pulley which articulates the talus with the inferior extremity of the tibia.

talus
The talus is the most elevated of the tarsal bones, located immediately below the inferior face of the tibia.

navicular
A flattened bone located in the internal border of the anterior row of the bones of the tarsus. Its internal face contains a projection called the navicular tuberosity which inserts some muscles

calcaneus
The largest of the bones of the tarsus. Its posterior face, which forms the posterior projection of the heel, is where the Achilles tendon is inserted.

cuboid
A bone located in front of the calcaneus with which it articulates posteriorly.

metatarsus
A set of five long bones that are united posteriorly with the bones of the tarsus and which are prolonged in their inferior part along each toe. They consist of a central, slightly curved and concave body with a triangular cross-section and two heavier ends.

cuneiform bones
The lateral, intermediate and lateral cuneiform bones are three small, wedge-shaped bones located in the anterior row of the tarsus.

phalanges of the toes
The phalanges constitute the skeleton of the fingers. Each finger consists of three phalanges called, from top to bottom, the first, second and third phalanges. The thumb, has only two. They are long bones, somewhat shorter than the phalanges of the fingers, although they have the same morphologic characteristics: a long body and two thicker extremes.

1st metatarsal

1st phalange

2nd phalange

3rd phalange

2nd metatarsal

3rd metatarsal

4th metatarsal

5th metatarsal

69

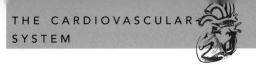

ARERIAL SYSTEM

▼ ANTERIOR GENERAL VIEW

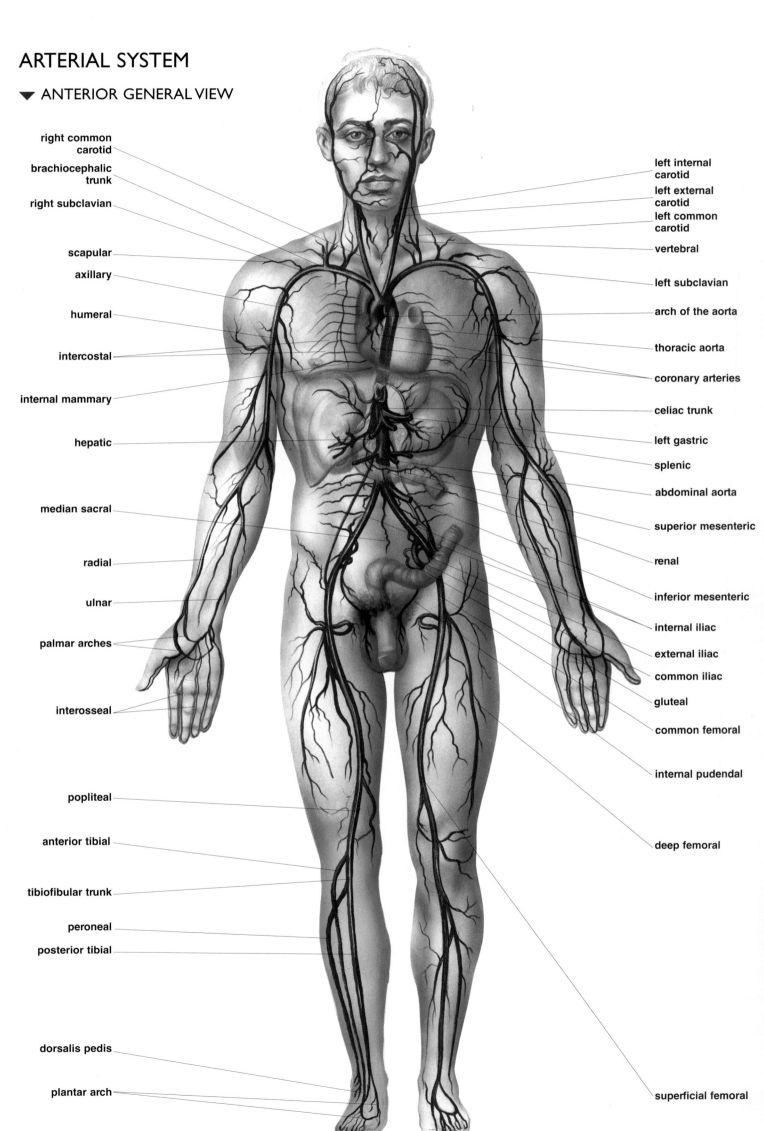

right common carotid

brachiocephalic trunk

right subclavian

scapular

axillary

humeral

intercostal

internal mammary

hepatic

median sacral

radial

ulnar

palmar arches

interosseal

popliteal

anterior tibial

tibiofibular trunk

peroneal

posterior tibial

dorsalis pedis

plantar arch

left internal carotid

left external carotid

left common carotid

vertebral

left subclavian

arch of the aorta

thoracic aorta

coronary arteries

celiac trunk

left gastric

splenic

abdominal aorta

superior mesenteric

renal

inferior mesenteric

internal iliac

external iliac

common iliac

gluteal

common femoral

internal pudendal

deep femoral

superficial femoral

70

VENOUS SYSTEM

▼ ANTERIOR GENERAL VIEW

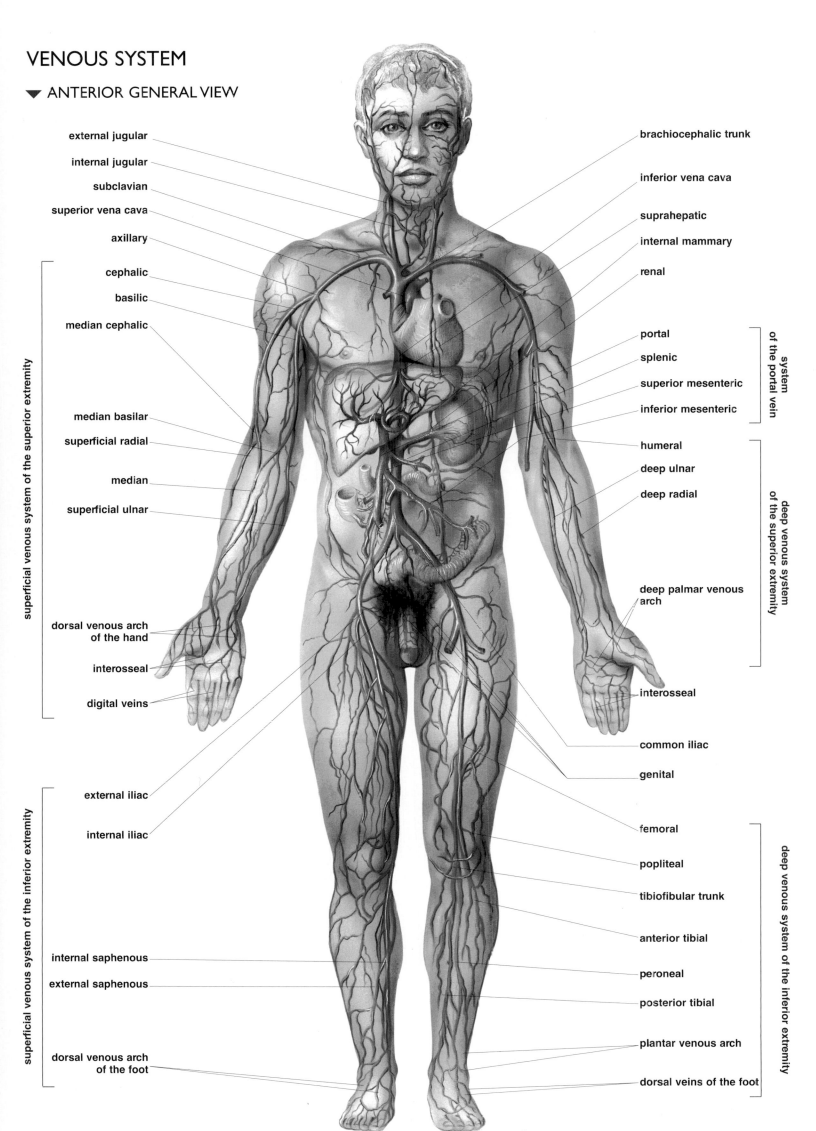

external jugular

internal jugular

subclavian

superior vena cava

axillary

cephalic

basilic

median cephalic

median basilar

superficial radial

median

superficial ulnar

dorsal venous arch of the hand

interosseal

digital veins

external iliac

internal iliac

internal saphenous

external saphenous

dorsal venous arch of the foot

superficial venous system of the superior extremity

superficial venous system of the inferior extremity

brachiocephalic trunk

inferior vena cava

suprahepatic

internal mammary

renal

portal

splenic

superior mesenteric

inferior mesenteric

system of the portal vein

humeral

deep ulnar

deep radial

deep palmar venous arch

interosseal

common iliac

genital

femoral

popliteal

tibiofibular trunk

anterior tibial

peroneal

posterior tibial

plantar venous arch

dorsal veins of the foot

deep venous system of the superior extremity

deep venous system of the inferior extremity

71

LYMPHATIC SYSTEM

▼ ANTERIOR GENERAL VIEW

LYMPHATIC SYSTEM

The lymphatic system is an accessory route
for the transport of fluids and the substances
they contain, especially proteins, coming
from the tissues of the body. It acts as a
complementary system of the arterial and
venous system and has its own network that
transports the lymphatic fluid or lymph,
to the blood.

cervical nodes
Groupings of lymph nodes that filter lymphatic
fluid from the head. They are located in the lateral
area of the neck, in the submaxillary region, the
area of the nape of the neck, the parotid region
and other cervical regions.

axillary nodes
A large group of nodes located under the skin
of the axilla. They filter the lymph from the
superior extremity before the lymph reaches
the venous blood.

lymph nodes
Thickenings of the lymphatic vessels
which are distributed throughout
the network. Their function is to filter
the lymph and to purify it of foreign
bodies. Although they exist in all areas
of the body, they are more common
in certain areas such as the inguinal,
axillary and cervical regions.

subclavian veins
The right and left subclavian veins channel
the blood from the axillary veins of the arms.
They pass below the clavicles to unite with
the jugular veins and then the superior vena
cava through the brachiocephalic venous
trunks. The subclavian veins receive the
great lymphatic vein (right) and the thoracic
lymph duct (left).

great lymphatic vein
A lymphatic duct located in the superior right
area of the thorax, which receives the
lymphatic vessels of the right half of the head,
thorax and right upper extremity. It joins the
right subclavian vein.

chyle cistern
A dilated saccular expansion in the
lower part of the thoracic duct located
posterior to the aorta into which
the two lumbar lymphatic trunks and
the intestinal trunk open.

inguinal ganglia
The inguinal area is especially rich in
lymph nodes which filter the lymph
coming from the inferior extremities.

thoracic lymph duct
A thick lymphatic duct that passes
through the abdomen and the thorax
vertically, parallel to the aorta, and
joins the left subclavian vein near its
union with the jugular vein. It collects
the lymph coming from the inferior
extremities, the intestine, the left half
of the thorax, the left arm and half
of the left side of the head.

Peyer's patch
Large aggregates of lymphoid tissue or
lymph nodes found in the small intestine,
which are part of the lymphatic system
that helps to fight infection.

lymphatic vessels
Ducts covering the entire body, closely
paralleling the venous system,
which collect the lymph coming from
the lymphatic capillaries.

lymphatic capillaries
Small ducts similar to the venous
capillaries, which begin in the sinuses
of all body tissues and collect
the lymph to transport it to the larger
lymphatic vessels.

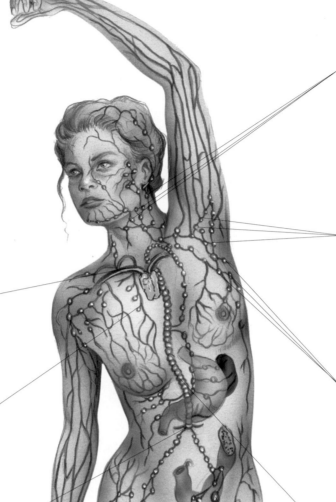

ARTERIES AND VEINS: INTERNAL STRUCTURE

ARTERIES

Blood vessels which carry the blood oxygenated in the lungs (arterial blood) to the tissues of the body. The arteries become smaller as they extend through the body, terminating in arterioles and arterial capillaries.

arterioles

Very small vessels which are the continuation of the successive ramifications of the arteries and which carry the blood to the arterial capillaries. Their tunic media is formed by a thin muscular lamina.

arterial capillaries

Microscopic ramifications of the arteries that carry the arterial blood to all parts of the body and facilitate the exchange of oxygenated blood and venous blood. At their extremes, they are united with the venous capillaries, which collect deoxygenated blood and transport it to the venous system.

VEINS

Blood vessels which carry the deoxygenated or venous blood, loaded with waste products, from the body tissues to the heart and lungs, where it will be oxygenated. The venous system includes the small venous capillaries, venules and medium-sized and large veins.

venules

Very small veins which join the medium and large-size veins and result from the progressive union of the different capillaries.

venous capillaries

A network of microscopic blood vessels which are the origin of the venous system. They collect the blood containing waste products from the different body tissues and transport it to the venules and larger veins.

73

tunica intima or endothelium

The tunica intima is the internal layer of the venous and arterial walls and rests on a layer of connective tissue.

tunica adventitia

The outer layer of the arteries and veins. It is formed of connective tissue and includes the nervous terminations and the blood capillaries that reach the arteries and veins.

subendothelial layer

A layer, located between the tunica intima and tunica media, which is well developed in large arteries and contains many elastic fibers, giving it a striated aspect.

tunica media

The middle layer of the three that form the arterial wall. It is composed of smooth, muscular fibers arranged concentrically and is especially abundant in the medium-sized arteries. In the larger arteries, this layer contains a large amount of elastic fibers that allow the artery to contract, expand and adapt in response to changes in the blood volume due to the contraction and relaxation of the heart.

tunica media

The intermediate layer of the vein wall which, unlike that of the arteries, has very few muscular fibers, but a large number of collagen fibers. Only the veins of the inferior half of the organism have a certain amount of muscular fibers which facilitate the ascent of the venous blood.

valves

Folds of the internal wall of the veins which are distributed throughout the system. Their function is to allow the blood to pass in the direction of the heart and prevent any reflux of venous blood.

HEART

▼ ANTERIOR SUPERFICIAL VIEW

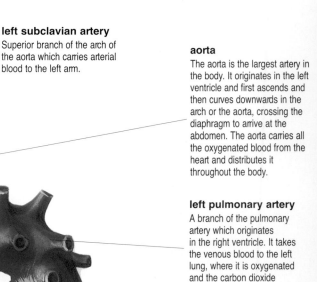

left common carotid artery
A large artery which originates in the arch of the aorta and carries the arterial blood to the left half of the head and the neck.

brachiocephalic arterial trunk
A thick, ascending, arterial branch that originates in the arch of the aorta. Its branches carry oxygenated blood to the right arm and right half of the neck and the head.

left subclavian artery
Superior branch of the arch of the aorta which carries arterial blood to the left arm.

aorta
The aorta is the largest artery in the body. It originates in the left ventricle and first ascends and then curves downwards in the arch or the aorta, crossing the diaphragm to arrive at the abdomen. The aorta carries all the oxygenated blood from the heart and distributes it throughout the body.

pericardium
A fibrous sac that covers all the heart, including the trunk of the aorta and the other large heart vessels (superior and inferior vena cava, pulmonary artery and veins).

left pulmonary artery
A branch of the pulmonary artery which originates in the right ventricle. It takes the venous blood to the left lung, where it is oxygenated and the carbon dioxide eliminated.

superior vena cava
A thick venous trunk which receives the veins of the arms and the head and carries their venous blood to the right atrium.

pulmonary artery
A thick trunk that originates in the right ventricle and divides into left and right branches which take the venous blood to the respective lungs.

right pulmonary artery
A branch of the pulmonary artery that originates in the right ventricle. It carries the deoxygenated blood to the right lung, where it is oxygenated and the carbon dioxide eliminated.

left atrium
The superior cavity of the heart, which receives the blood oxygenated in the lungs from the pulmonary veins. It has relatively thin walls

right atrium
A cavity with thin walls which collects the venous blood coming from the superior and inferior vena cava.

left coronary artery
A branch of the aorta which descends bordering the left side of the heart through the sulcus between the atrium and the left ventricle. It has branches which go to the interventricular area and the left side of the heart.

atrioventricular or coronary sulcus
A fold or sulcus that separates the two atriums and two ventricles of the heart.

great cardiac vein
A vein that occupies the atrioventricular sulcus collecting the venous blood of the branches of the left side of the body and transporting it to the coronary sinus.

small cardiac vein
A vein that travels through the atrioventricular sulcus collecting the blood from the venous branches of the right side of the body and carries it to the coronary sinus.

right coronary artery
A branch of the aorta originating in the anterior face of the heart which crosses the right side of the heart through the atrioventricular sulcus and carries arterial blood to the posterior face of the heart. It has branches which go to the right border and posterior face of the heart and another interventricular branch.

coronary artery of the right border
An arterial branch originating in the right coronary artery which carries oxygenated blood to the right side of the heart.

right ventricle
The inferior cavity of the heart, which receives the venous blood from the right atrium and it sends it through the pulmonary artery to the lungs. It has thick muscular walls.

interventricular coronary artery
A branch of the left coronary artery that descends the anterior face of the heart in the area of the interventricular septum.

left ventricle
A large cavity that receives the oxygenated blood from the left atrium and, by means of powerful contractions of its thick walls, sends it through the aorta to all parts of the body.

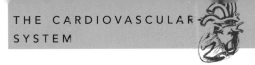

HEART

▼ INTERNAL VIEW

aorta
A large artery which transports all the arterial blood from the left ventricle to distribute it throughout the body.

aortic valve
A valve located in the communicating orifice between the left ventricle and the aorta. Its opening in systole (ventricular contraction) allows the oxygenated blood to reach the aorta and its closing in diastole (ventricular relaxation) closes this circulation and prevents the blood returning from the aorta to the ventricle.

left pulmonary vein
One of the four blood vessels, two from the right lung and two from the left, which carry the oxygenated blood from the lungs to the left atrium.

left atrium
One of the two superior cavities of the heart, formed by thin walls whose superior part contains the orifices of the four pulmonary veins which deliver blood from the lungs to the heart. The left atrium communicates with the left ventricle through the mitral valve.

superior and inferior vena cava
The superior and inferior vena cava are joined at the height of the right atrium in which they deposit the venous blood from the entire body.

coronary arteries
The arterial system which runs through all the heart.

right atrium
A cavity with thin walls located in the superior area of the heart. It receives all the venous blood from the body from the venae cavae. It communicates with the immediately inferior cavity, the right ventricle, through the tricuspid valve.

mitral valve
A valvular system equipped with two valves that separates the left atrium from the left ventricle. Its opening and closing permits the blood to flow from the atrium to the ventricle, but not from the ventricle to the atrium.

pericardium
The pericardium is a fluid-filled sac that surrounds the heart and the proximal terminates of the aorta, vena cava and the pulmonary artery. It is divided into three layers: the fibrous, visceral and parietal pericardium.

myocardium
The walls of the heart are composed mainly of cardiac muscle cells called myocardium, which enable the heart to contract and relax. The myocardium is very thick around the ventricles and thinner around the atria. It contains the nerves which mark the rhythmic movement of

tricuspid valve
A valvular system that separates the right atrium from the right ventricle. It opens to permit blood to flow from the atrium to the ventricle and contracts to prevent blood flowing from ventricle to atrium. The opening and closing of the heart valves makes a click, the sound of the heart beats which can be heard through a stethoscope.

endocardium
The innermost layer of the wall of the heart. It is a thin layer of connective tissue made of endothelial cells which covers all the structures of the heart.

papillary muscles
Muscular columns that serve as a prolongation of the mitral and tricuspid valves and which are attached to the ventricular walls, facilitating the movement of the valve by their contractions.

right ventricle
A large cavity located in the inferior part of the heart. It has thick muscular walls and its function is to store the venous blood coming from the right atrium and to send it, by means of abrupt contractions, towards the lungs, where it is oxygenated and purified.

interventricular septum
A thick wall of powerful muscular tissue that completely separates the two ventricular cavities and is responsible for their movements. The superior part is thinner and more fibrous.

left ventricle
A cavity with thick walls that receives the oxygenated blood from the left atrium and, by means of strong contractions, expels it towards the arterial system, which carries it through the body. As the effort of the left ventricle is greater than that of the right, the muscular walls are thicker.

75

ARTERIAL SYSTEM: AORTA

AORTA

The largest artery, which carries the blood flow to all parts of the human body. It originates in the right ventricle and after ascending forms a descending curve called the arch of the aorta, crosses the thorax as the thoracic aorta. It crosses the diaphragm, to become the abdominal aorta, which, near the pelvic cavity divides into the common iliac arteries which irrigate the inferior extremities.

right brachiocephalic trunk

A thick arterial trunk that originates in the right part of the arch of the aorta and rapidly bifurcates into an ascending branch, which carries the blood to the head (right common carotid artery), and a horizontal branch, which carries blood to the right superior extremity (subclavian artery).

left common carotid artery

Unlike the right side of the body, the left side does not posses a common brachiocephalic trunk. The left common carotid artery originates directly in the arch of the aorta and ascends through the neck to the left part of the head, bifurcating into the internal and external carotid arteries.

renal arteries

Two arteries that branch laterally and horizontally from the aorta and go to the kidneys.

hepatic artery

The hepatic artery is a branch of the celiac trunk that goes to the liver, which it irrigates.

genital arteries

Two arteries that descend to the testicles in men (testicular arteries) or the ovaries in women (ovarian arteries).

lumbar arteries

Five arteries that branch perpendicularly from the abdominal aorta and irrigate the muscles and other structures of the walls of the abdominal cavity.

abdominal aorta

The name given to the section of the aorta that crosses the diaphragm and enters the abdominal cavity. The first branches go to the diaphragm and later branches include the celiac trunk, the renal arteries and the superior and inferior mesenteric arteries. The abdominal artery terminates in a bifurcation into the common iliac arteries.

right subclavian artery

The horizontal branch of the bifurcation of the brachiocephalic trunk. It originates all the arteries that irrigate the right superior extremity.

right common carotid artery

The ascending branch of the bifurcation of the brachiocephalic trunk. It branches into the internal and external right carotid arteries which irrigate the intra and extracranial structures of the right side of the head.

left vertebral artery

The ascending branch of the left subclavian artery which ascends through the neck parallel to the cervical vertebral column and enters the skull through the foramen magna, originating the arterial network that irrigates the posterior part of the brain and cerebellum.

inferior thyroid artery

The inferior thyroid artery originates in the left subclavian artery and ascends through the neck. It has branches that go to the esophagus, trachea, larynx and thyroid.

left subclavian artery

The left subclavian artery carries arterial blood to the left superior extremity. It originates directly in the arch of the aorta, unlike the right side of the body where a common brachiocephalic trunk supplies the blood to the head and the right superior extremity.

arch of the aorta

On leaving the left ventricle, the aorta first ascends and then immediately turns left, forming a descending arch. The branches of this region of the aorta supply all the arterial blood to the head and superior extremities.

thoracic aorta

The portion of the aorta that crosses the thoracic cavity vertically from the arch of the aorta to the diaphragm. It has branches that go to the esophagus, the bronchi, the mediastinum and the intercostal areas.

intercostal arteries

Perpendicular branches of the thoracic aorta which enter the intercostal spaces laterally. There are twelve intercostal arteries plus posterior branches to the vertebrae and anterior branches to the intercostal muscles, pleura and ribs.

celiac trunk

A thick arterial trunk that originates in the front part of the abdominal aorta. It supplies arterial blood to the liver, stomach and spleen through the hepatic, gastric coronary and splenic arteries.

splenic artery

The left branch of the celiac trunk which irrigates the spleen.

gastric coronary artery or stomachic

A branch of the celiac trunk that irrigates the internal area of the stomach.

superior mesenteric artery

An artery that appears below the origin of the celiac trunk and also in the anterior face of the abdominal aorta. It irrigates the small intestine, a part of the pancreas and the mesentery as well as the right portion of the large intestine.

inferior mesenteric artery

An artery that irrigates the left part of the large intestine, from the middle of the transverse colon to the rectum through successive ramifications (colic, sigmoid, hemorrhoidal arteries).

common iliac arteries

The common iliac arteries originate in the final bifurcation of the abdominal aorta and descend obliquely towards the inferior extremities. At the height of the sacroiliac union, they bifurcate into internal and external branches. Other branches go to some of the pelvic muscles of the abdominal cavity.

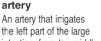
ARTERIAL SYSTEM: ABDOMEN

pancreatoduodenal artery
An arterial branch that originates in the superior mesenteric artery and goes to the duodenum and the left part of the pancreas.

median colic artery
The median colic artery originates in the superior mesenteric artery and irrigates the transverse colon. It forms an arterial network which unites with the branches of the inferior mesenteric artery.

superior mesenteric artery
An artery that appears below the origin of the celiac trunk and also in the anterior face of the abdominal aorta. It irrigates the small intestine, a part of the pancreas and the mesentery and the right portion of the large intestine.

inferior pancreatic artery
An arterial branch of the superior mesenteric artery that irrigates the inferior border of the pancreas.

abdominal aorta
The name given to the section of the aorta that crosses the diaphragm and enters the abdominal cavity. The first branches go to the diaphragm and later branches include the celiac trunk, the renal arteries and the superior and inferior mesenteric arteries. The abdominal artery terminates in a bifurcation into the common iliac arteries.

right colic artery
The right colic artery originates in the right side of the superior mesenteric artery and divides into branches which go to the ascending colon.

inferior mesenteric artery
An artery that irrigates the left part of the large intestine, from the middle of the transverse colon to the rectum, through successive ramifications (colic, sigmoid and hemorrhoidal arteries).

ileocolic artery
A branch of the superior mesenteric artery which, like the right colic artery, goes to the ascending colon and also has branches to the final part of the ileum.

left colic artery
The left colic artery originates in the inferior mesenteric artery and goes to the descending colon. It has some branches that go to part of the transverse colon, uniting with branches of the superior mesenteric artery.

jejunal and ileal arteries
Arterial branches that go to the ileum and jejunum through a series of arches which cross the mesentery.

sigmoid arteries
The sigmoid arteries form a network that originates in the inferior mesenteric artery and goes to the final part of the large intestine, the sigmoid colon, rectum and anus

common iliac artery
The common iliac arteries originate in the final bifurcation of the abdominal aorta and descend obliquely towards the inferior extremities. At the height of the sacroiliac union, they bifurcate into internal and external branches. Other branches go to some of the pelvic muscles of the abdominal cavity.

hemorrhoidal arteries
Arteries originating in the inferior mesenteric artery that go to the final part of the anus and constitute the hemorrhoidal plexus. A part of this area is irrigated by arteries coming from the internal iliac artery.

77

external iliac artery
The most external branch of the two into which the common iliac artery is divided. It crosses the pelvic cavity obliquely to reach the inguinal area and originates the arteries of the inferior extremity. It has branches to the ureter and abdomen and a branch called the epigastric artery that ascends by the anterior wall of the abdomen.

internal iliac artery
The common iliac artery bifurcates into two branches: the internal and external iliac arteries. The internal iliac also called the hypogastric artery, goes to the viscera of the pelvic cavity such as the bladder and uterus (intrapelvic branches) and to the external genitals and the internal part of the thigh (extrapelvic branches).

ARTERIAL SYSTEM: HEAD AND NECK

superficial temporal artery
One of the branches into which the external carotid artery bifurcates at the height of the mandibular joint. It ascends through the temporal area and has branches that go to the face, the mandibular joint, the auricular pavilion and the orbital area, bifurcating into frontal and parietal branches.

parietal artery
The posterior branch of the bifurcation of the superficial temporal bone. It has numerous branches which go to the parietal area of the skull.

frontal artery
The anterior branch of the bifurcation of the superficial temporal artery. It goes to the forehead where it has numerous branches.

internal maxillary artery
The internal maxillary artery originates in the terminal bifurcation of the external carotid, passes under the zygomatic arch and enters the skull through the sphenopalatine foramen and goes to the nasal septum and conchae. It has numerous branches that go to the tympanum, the temporal fossa, the dental and buccal area, the palate, the masseter muscle and the pharynx. In the skull, it also has meningeal branches.

posterior auricular artery
An artery that originates in the posterior face of the internal carotid artery and has branches going to the parotid gland. It terminates in a bifurcation with branches going to the mastoid region and the auricular pavilion.

occipital artery
The occipital artery originates in the posterointernal face of the external carotid artery and goes posteriorly to the occipital area.

vertebral artery
A branch of the subclavian artery which ascends posteriorly to irrigate the musculature of the cervical vertebral area and enters the cranium through the foramen magna. It has branches going to the meninges, medulla oblongata and cerebellum.

facial artery
A branch of the external carotid artery that borders the mandible and goes to the face, passing close to the commissure of the lips and terminating in the internal angle of the eye. It has submentonian branches and other facial branches which go to the masseter muscle, the superior and inferior lips and the area of the ala of the nose.

internal carotid artery
The internal bifurcation of the common carotid artery. It ascends and enters the cranium through the carotid foramen. The internal carotid has multiple branches that irrigate the brain, the eyeball (ophthalmic artery) and other intracranial structures.

lingual artery
The anterior branch of the external carotid artery, which passes below the mandible and goes to irrigate the muscles of the tongue.

subclavian artery
The external branch of the bifurcation of the brachiocephalic trunk. It goes to the arm and originates all the arterial circulation of the superior extremity.

external carotid artery
The external carotid artery originates in the bifurcation of the common carotid artery and goes to the area of the mandibular joint, where it generates two terminal branches that go to the maxilla, the temporal area and the auricular area. It also has collateral branches that go to the thyroid, larynx and tongue.

right brachiocephalic trunk
A thick arterial trunk that originates in the highest part of the arch of the aorta and immediately bifurcates into the common carotid artery, which supplies almost all the arterial irrigation of the head, and the subclavian artery, which originates all the arteries of the superior extremity.

common carotid artery
A branch of the brachiocephalic trunk that ascends following the lateral border of the neck and carries the blood to one side of the head. It branches into the external and internal carotid arteries.

superior thyroid artery
Shortly after the external carotid artery arises from the common carotid, it gives rise to the superior thyroid artery, which descends and irrigates the thyroid. Its branches go to the larynx, the sternocleidomastoid muscle and the muscles of inferior area of the hyoid bone.

ARTERIAL SYSTEM: CRANIAL BASE

anterior cerebral artery
The anterior cerebral artery is a branch of the internal carotid. It runs above the optic nerve to the corpus callosum and is joined by the anterior communicating artery. It has branches which pass to the anterior part of the internal capsule and basal nuclei.

middle cerebral artery
The middle cerebral artery originates in the internal carotid and goes laterally to the external face of the brain. It irrigates, both superficially and deeply, part of the frontal lobe, the temporal lobe and the parietal lobe of the brain.

posterior cerebral artery
The posterior cerebral artery originates in the anterior bifurcation of the basilar artery and, after passing the cerebral peduncle, goes laterally and posteriorly to the inferior surface of the occipital and temporal lobes, with deep branches to the interior of these areas.

basilar artery
The basilar artery originates in the union of the two vertebral arteries and goes anteriorly to bifurcate in the two posterior cerebral arteries.

vertebral artery
The artery which carries blood to the posterior part of the intracranial structures. It originates in the subclavian artery and, after ascending through the neck, it enters in the skull through the foramen magna and goes to the central area where the two vertebral arteries are united to form the basilar artery, which branches into the median and posterior cerebral arteries and the anterior spinal artery.

anterior spinal artery
The anterior spinal artery is one of the major arteries to the spinal cord and is formed from branches of the vertebral arteries which unite in the median fissure of the spinal cord. It runs the length of the spinal cord.

polygon of Willis
A polygonal figure formed by the union of different arteries of the base of the cranium. It contains the optical chiasm and the hypophysial stem which unites the hypophysis with the brain. The sides of the polygon are formed by the anterior cerebral arteries, united by the anterior communicating artery, the posterior cerebral arteries and the posterior communicating arteries.

internal carotid artery
An artery which provides a large part of the cerebral circulation. It enters the skull from the neck through the carotid foramen. Its main branches are the ophthalmic artery, the anterior cerebral artery and the middle cerebral artery.

posterior communicating artery
An arterial branch of the internal carotid artery which terminates in the posterior cerebral artery, communicating the two arterial systems that irrigate the brain, the internal carotid and vertebral arteries.

superior cerebellar artery
It originates in the basilar artery and goes to the superior surface of the cerebellum.

posterior inferior cerebellar artery
A branch of the vertebral artery which winds through the inferior face of the cerebellum, irrigating it.

anterior inferior cerebellar artery
This artery originates in the basilar artery, close to the union of the basilar and vertebral arteries, and goes to the anteroinferior area of the cerebellum.

79

ARTERIAL SYSTEM: SHOULDER AND ARM

anterior humeral circumflex artery
A thin arterial branch that originates in the axillary artery and passes in front of the humerus. It irrigates the shoulder joint, the deltoid muscle, the biceps and other muscles.

humeral or brachial artery
The humeral artery is a continuation of the axillary artery, which goes from the axilla to the elbow, where it divides into an external or radial branch and an internal or cubital branch. It also has branches that go to the muscles of the arm, nutrient branches to the humerus and several other collateral branches.

radial collateral artery
A large artery that originates in the humeral artery. It passes behind the humerus and descends the posterior face of the arm to its external area, passing the elbow and uniting with the anterior radial recurrent artery. It has various branches which go to the triceps muscle.

radial recurrent artery
An artery that branches from the radial artery shortly after its origin, following a retrograde or ascending path to unite with the deep humeral artery. It irrigates the epicondyle muscles of the external part of the elbow and forearm.

radial artery
An artery that originates in the bifurcation of the humeral artery in the flexure of the elbow. It follows the external border of the forearm to the carpus. It has branches which go to the muscles of the anterior part of the arm and the carpal area.

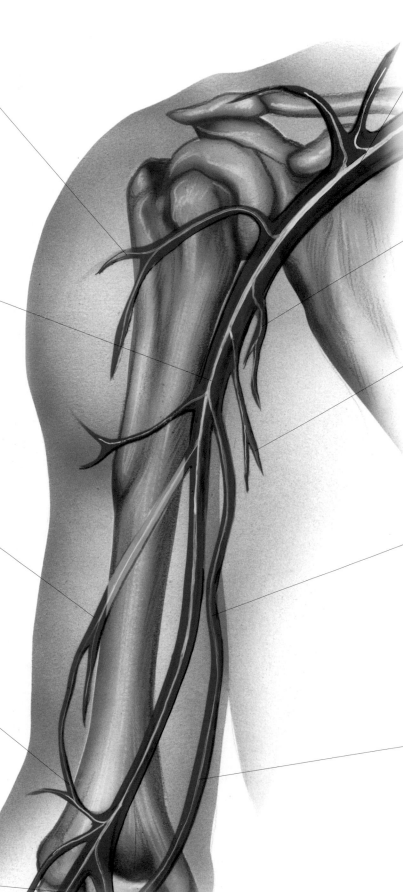

axillary artery
When the subclavian artery passes the clavicle, it becomes the axillary artery which crosses the axilla to the arm. It has mammary, thoracic and chest and scapular branches that remain in the region of the thorax and the shoulder, and circumflex branches which go to the arm.

posterior humeral circumflex artery
A branch of the axillary artery which passes behind the humerus, carrying blood to the triceps, some shoulder muscles and the deltoid muscles.

superior ulnar collateral artery
An artery which originates in the humeral artery and descends along the internal face of the arm to reach the elbow, where it is united with the anterior ulnar recurrent artery. It has small branches that go to the internal part or vastus medialis of the triceps.

inferior ulnar collateral artery
A thin artery which branches from the humeral artery and goes to the elbow, where it divides into anterior and posterior branches which are united with the anterior and posterior recurrent ulnar arteries, respectively. It irrigates the muscles of the internal area of the elbow and forearm or the epitrochlear muscles.

anterior recurrent ulnar artery
An artery that branches off from the ulnar artery shortly after its origin, and follows an ascending path to unite with the anterior branch of the inferior ulnar collateral artery. There is also a posterior ulnar recurrent artery that follows a parallel path along the upper surface of the elbow.

ulnar artery
The internal branch of the bifurcation of the humeral artery. It passes along the internal border of the forearm to the palmar area. It irrigates the posterior area of the forearm and reaches the interosseal region of the hand.

ARTERIAL SYSTEM: FOREARM AND HAND

radial recurrent artery
An artery that branches from the radial artery shortly after its origin, and follows a retrograde or ascending path to unite with the deep humeral artery. It irrigates the epicondyle muscles of the external part of the elbow and forearm.

radial artery
An artery that originates in the bifurcation of the humeral artery in the flexure of the elbow. It follows the external border of the forearm to the carpus. It has branches which go to the muscles of the external forearm and the carpal area.

palmar branch of radial artery
A branch of the radial artery that originates near the wrist and goes to the palm of the hand, where it unites with the termination of the ulnar artery to form the superficial palmar arch.

artery of the thumb
A branch of the radial artery which goes to the thumb, crossing its posterior face.

common digital palmar arteries
Arteries that branch off from the superficial palmar arch and follow a parallel path to the metacarpal bones, reaching the roots of the fingers, where they are transformed into the proper palmar digital arteries.

proper palmar digital arteries
Arteries that originate in the common digital palmar arteries. There are two for each finger and they run along the external and internal borders of the fingers.

humeral or brachial artery
An artery proper to the arm which is a continuation of the axillary artery, going from the axilla to the elbow, where it divides into an external branch, the radial artery and an internal branch, the ulnar artery. It has branches to the muscles of the arm, nutritional branches to the humerus and several other collateral branches.

anterior ulnar recurrent artery
An artery that branches off from the ulnar artery shortly after its origin and follows a retrograde or ascending path to unite with the anterior branch of the inferior ulnar collateral artery. There is also a posterior ulnar recurrent artery that follows a parallel path along the dorsal face of the elbow.

interosseal artery
A branch of the ulnar artery which goes between the ulna and the radius. It bifurcates into an anterior branch, which remains in the anterior face of the forearm, and a posterior branch, which crosses the ligament uniting both bones and passes to the dorsal face. The two branches carry blood to the muscles of the forearm.

ulnar artery
The internal branch of the bifurcation of the humeral artery. It passes along the internal border of the forearm to the palmar area, with branches that irrigate the muscles of these areas.

palmar branch of ulnar artery
A branch of the ulnar artery which goes to the palm of the hand where it is united with the termination of the radial artery, giving rise to the deep palmar arch. It has small branches which go to the muscles of the hypothenar eminence.

superficial palmar arch
Formed by the union of the palmar branch of the radial artery, a branch of the radial artery, with the termination of the ulnar artery. The arch gives rise to digital arteries which carry blood to the fingers.

deep palmar arch
The result of the union of the palmar ulnar artery, a branch of the ulnar artery, with the termination of the radial artery. Its branches are the interosseal arteries which unite with the digital arteries.

81

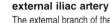

ARTERIAL SYSTEM: THIGH

crural arch
A fibrous ligament that extends from the anterosuperior iliac spine to the pubis. The arteries, veins and nerves of the inferior extremity pass under the arch to the thigh.

common femoral artery
A continuation of the external iliac artery, which begins in the inguinal area and goes to the thigh, where it bifurcates into a superficial femoral artery and a deep branch. It emits branches to the genital area called the pudendal branches and others to the walls of the abdomen.

external or **anterior circumflex**
A branch of the deep femoral artery that extends outwards and irrigates the muscles of the thigh area and the hip joint.

deep femoral artery
A branch of the common femoral artery which goes between the muscles of the thigh to become posterior. It emits ramifications to the head of the femur and the quadriceps, adductor and flexor muscles.

popliteal artery
A continuation of the femoral artery that begins at the top of the popliteal fossa, which it crosses vertically. It sends branches to the knee joint and the gastrocnemius muscle and divides into the anterior tibial artery and the tibiofibular trunk.

external iliac artery
The external branch of the two into which the common iliac artery divides. It crosses the pelvic cavity obliquely to reach the inguinal area and has branches to the ureter and abdomen and a branch called the epigastric artery that ascends the anterior wall of the abdomen.

internal iliac artery
The internal iliac artery, also called the hypogastric artery, goes to the organs of the pelvic cavity such as the bladder and uterus (intrapelvic branches) and to the external genitals and the internal part of the thigh (extrapelvic branches).

internal or **posterior circumflex artery**
It originates in the posterior face of the deep femoral artery and passes behind the humerus going to the inferior part of the gluteal region.

superficial femoral artery
From its origin in the bifurcation of the common femoral artery, the superficial femoral crosses the internal face of the thigh and becomes posterior when it reaches the popliteal fossa, becoming the popliteal artery. Its branches go to the quadriceps muscle.

genicular artery
An artery which originates in the superficial femoral artery and descends the internal border of the thigh to bifurcate into a superficial or saphenous branch and another deep or articular branch.

82

ARTERIAL SYSTEM: LEG AND FOOT

popliteal artery
A continuation of the femoral artery that begins at the height of the popliteal fossa, which it crosses vertically. It sends branches to the knee joint and the gastrocnemius muscle and divides into the anterior tibial artery and the tibiofibular trunk.

anterior tibial recurrent artery
A branch of the anterior tibial artery which irrigates the area of the knee.

anterior tibial artery
A branch of the popliteal artery which becomes anterior after crossing the interosseal space located between the tibia and the fibula. It continues along the external part of the leg, crosses the ankle and goes to the back of the foot. It has branches which go to the peroneal muscles, the areas of the internal and lateral malleoli, and a recurrent or retrograde branch which goes to the knee.

dorsalis pedis artery
A continuation of the anterior tibial artery which begins at the dorsal area of the foot. It irrigates the tarsus and metatarsus and emits branches that are united with the plantar arteries, forming the plantar arch.

external malleolar artery
A branch of the anterior tibial artery which irrigates the area of the lateral malleolus.

interosseal arteries or metatarsals
Arteries that originate in the plantar arch and continue in the interosseal spaces of the four last metatarsals, emitting branches that reach the toes.

genicular artery
An artery which originates in the superficial femoral artery, and descends the internal border of the thigh to bifurcate into a superficial or saphenous branch and another deep or articular branch.

tibiofibular trunk
A short section of artery which results from the bifurcation of the popliteal artery and goes to the posterior area of the leg, where it divides the posterior tibial artery and peroneal artery.

posterior tibial artery
A branch of the internal bifurcation of the tibiofibular trunk. It descends the posterointernal side of the leg, emitting branches that go to the muscles of the area and the tibia. It crosses the ankle joint and gives rise to the plantar arteries of the foot.

peroneal artery
It originates in the tibiofibular trunk and goes to the posteroexternal area of the leg where it irrigates the muscles of the area and the fibula and reaches the heel of the foot.

internal malleolar artery
A branch of the anterior tibial artery which irrigates the area of the medial malleolus.

internal and external plantar arteries
Terminal branches of the posterior tibial artery that cross the internal and external borders of the foot and are united to form the plantar arch.

plantar arch
An arch, formed by the union of the plantar arteries of the posterior tibial artery and the terminal branches of the dorsalis pedis artery, which crosses the sole of the foot. The arch gives rise to the interosseal arteries and those that irrigate the toes.

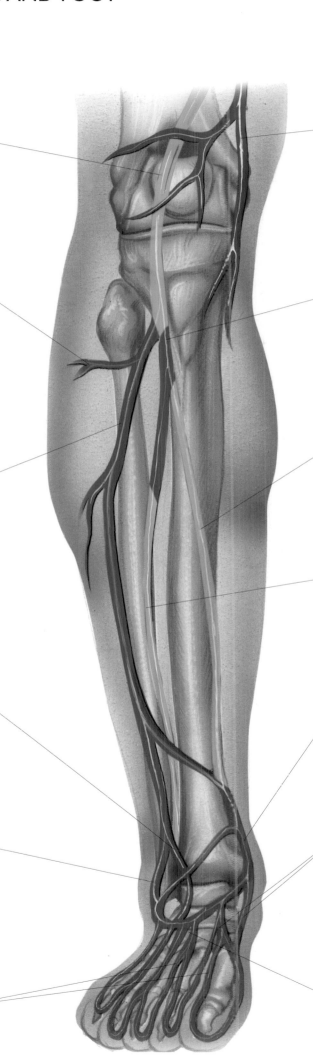

83

VENOUS SYSTEM: SUPERFICIAL VEINS OF THE FOOT AND LEG

VENOUS SYSTEM OF THE INFERIOR EXTREMITY

The superior and inferior extremities are equipped with twin venous systems, one deep and the other superficial. The deep venous system is a system parallel to the arterial system, with identical names and trajectory, although there are two veins for each artery. For these reasons, it is not considered necessary to illustrate it. The superficial venous system, on the contrary, has a different trajectory and nomenclature and runs up the leg in more superficial areas, although it terminates in the deep venous system.

subcutaneous veins of the anterior face of the knee
A dense venous network that ascends the leg just below the skin and carries the venous blood of this area to the great saphenous vein.

small saphenous vein
A vein that originates in the external part of the dorsal venous arch of the foot and, after passing behind the lateral malleolus of the ankle, ascends the posterior area of the leg and reaches the thigh, where it unites with the popliteal vein of the deep venous system. It also has a branch which terminates in the great saphenous vein.

dorsal venous arch of the foot
A venous network located in the superficial area of the back of the foot, formed by the confluence of the digital veins and some plantar veins. It ascends internally to the great saphenous vein and externally to the external saphenous.

veins of the toes
Small veins that originate in the extremities of the toes, basically in the dorsal face, and which carry the venous blood of this area to the dorsal arch of the leg.

opening of the small saphenous vein in the popliteal vein
When it reaches the popliteal fossa in the posterior part of the knee, the small saphenous vein terminates in the popliteal vein, part of the deep venous system of the inferior extremity.

communication between the external and great saphenous veins
When it passes the knee posteriorly, the small saphenous vein terminates in the popliteal vein of the deep venous system, but it also has a communicating branch which unites it with the great saphenous vein.

subcutaneous veins of the anterior face of the leg
A dense venous network that crosses the knee and the leg just below the skin and carries venous blood to the great saphenous vein.

great saphenous vein
A vein that originates in the internal part of the dorsal venous arch of the foot and, after passing in front of the medial malleolus of the ankle, ascends the leg, collecting the venous blood of the subcutaneous venous network of the anterior and internal part of the leg. After passing the knee, it reaches the thigh, crossing it superficially in its anterointernal face to terminate in the final portion of the femoral vein.

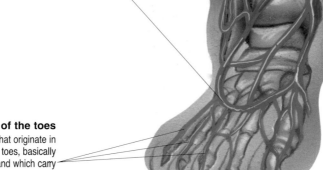

VENOUS SYSTEM: SUPERFICIAL VEINS OF THE THIGH

**VENOUS SYSTEM
OF THE INFERIOR EXTREMITY**

The superior and inferior extremities are equipped with twin venous systems, one deep and the other superficial. The deep venous system is a system parallel to the arterial system, with identical names and trajectory, although there are two veins for each artery. For these reasons, it is not considered necessary to illustrate it. The superficial venous system, on the contrary, has a different trajectory and nomenclature and runs up the thigh in more superficial areas, although it terminates in the deep venous system.

circumflex iliac vein

A vein that follows a parallel path to the artery of the same name and collects the venous blood from the superficial area of the lateral walls of the abdomen. It terminates in the femoral vein.

femoral vein

The superficial and deep venous systems of the leg and thigh converge in the femoral vein, which ascends the thigh posteriorly and receives the great saphenous vein in the inguinal area. After crossing the crural arch it continues as the external iliac vein.

subcutaneous veins of the anterior face of the thigh

A dense venous network that crosses the knee and the leg just below the skin and carry the venous blood of this area to the great saphenous vein.

anamostosis of the small and great saphenous veins

When it passes the knee posteriorly, the small saphenous vein terminates in the popliteal vein of the deep venous system, but it also has a communicating branch which unites it with the great saphenous vein.

small saphenous vein

A vein that originates in the external dorsal part of the foot and after passing behind the lateral malleolus of the ankle, reaches the thigh, where it unites with the popliteal vein of the deep venous system. It also has a branch which terminates in the great saphenous vein.

external iliac vein

A thick vein which is a continuation of the femoral vein. It receives the venous blood from the inferior extremity and carries it to the inferior vena cava, where the internal iliac vein also terminates.

crural arch

A fibrous ligament that extends obliquely from the anterosuperior iliac spine to the pubis and represents the limit between the pelvic and femoral regions. The arteries, veins and nerves of the inferior extremity pass under the crural arch to the thigh.

pudendal veins

Veins that collect the venous blood from a part of the genitals and terminate in the great saphenous vein near its union with the femoral vein.

accessory saphenous vein

The accessory saphenous collects the venous blood from the posterior part of the thigh. It terminates in the superior part of the great saphenous vein.

great saphenous vein

A vein that originates in the internal part of the dorsal venous arch of the foot and, after passing in front of the medial malleolus of the ankle, ascends the leg, collecting the venous blood of the subcutaneous venous network of the anterior and internal part of the leg. After passing the knee, it reaches the thigh, crossing it superficially in its internal face to terminate in the final portion of the femoral vein.

85

VENOUS SYSTEM: SUPERFICIAL VEINS OF THE HAND AND FOREARM

VENOUS SYSTEM OF THE SUPERIOR EXTREMITY

The superior and inferior extremities are equipped with twin venous systems, one deep and the other superficial. The deep venous system is a system parallel to the arterial system, with identical names and trajectory, although there are two veins for each artery. For these reasons, it is not considered necessary to illustrate it. The superficial venous system, on the contrary, has a different trajectory and nomenclature and runs up the forearm in more superficial areas, although it terminates in the deep venous system.

cephalic vein

The cephalic vein originates in the meeting point between the median cephalic vein and the superficial radial vein. It ascends the arm in its external superficial area and terminates in the axillary vein.

median cephalic vein

One of the veins ascends the anterior face of the flexure of the elbow, from the bifurcation of the median vein to unite with the superficial radial vein to form the cephalic vein.

superficial radial vein

The superficial radial crosses the forearm superficially, firstly posteriorly and subsequently externally. It collects the venous blood from the forearm and also from the external dorsal part of the hand. It unites with the median cephalic vein to form the cephalic vein.

median vein

Vein that ascends the anterior face of the forearm, from the palmar area of the hand to the elbow, where it bifurcates into the median cephalic and median basilic veins. It is joined by various venous branches coming from the anterior face of the forearm.

dorsal venous arch of the hand

The dorsal arch forms the network of veins of the back of the hand, which it collects from the interosseal veins and distributes by ramifications to the radial and ulnar veins.

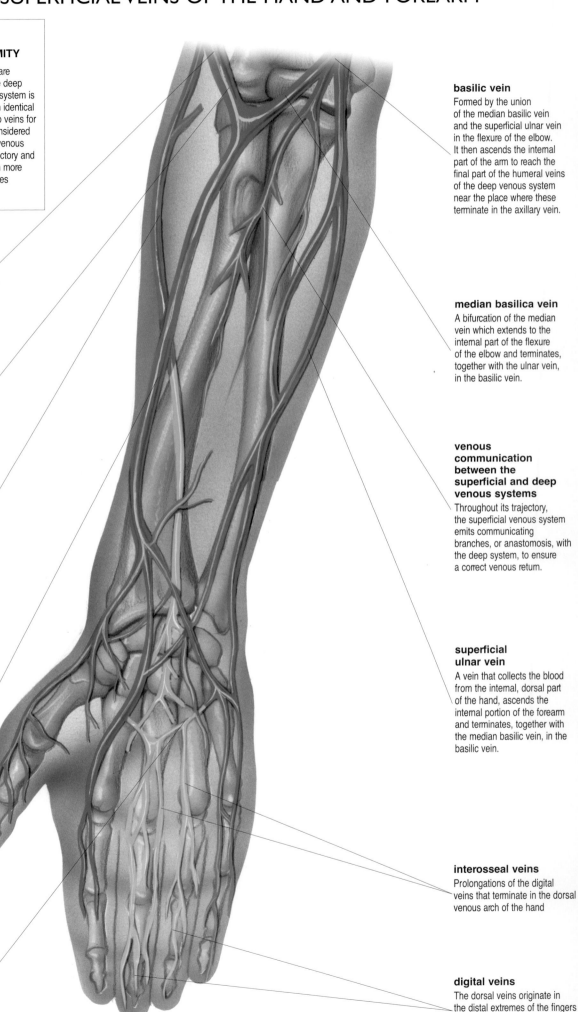

basilic vein

Formed by the union of the median basilic vein and the superficial ulnar vein in the flexure of the elbow. It then ascends the internal part of the arm to reach the final part of the humeral veins of the deep venous system near the place where these terminate in the axillary vein.

median basilica vein

A bifurcation of the median vein which extends to the internal part of the flexure of the elbow and terminates, together with the ulnar vein, in the basilic vein.

venous communication between the superficial and deep venous systems

Throughout its trajectory, the superficial venous system emits communicating branches, or anastomosis, with the deep system, to ensure a correct venous return.

superficial ulnar vein

A vein that collects the blood from the internal, dorsal part of the hand, ascends the internal portion of the forearm and terminates, together with the median basilic vein, in the basilic vein.

interosseal veins

Prolongations of the digital veins that terminate in the dorsal venous arch of the hand

digital veins

The dorsal veins originate in the distal extremes of the fingers and carry the venous blood to the interosseal veins.

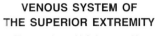
VENOUS SYSTEM: SUPERFICIAL VEINS OF THE ARM AND SHOULDER

axillary vein

The vein through which all the blood from the superficial and deep venous systems of the superior extremity is carried to the subclavian vein. It originates in the axillary region from the union of the cephalic and basilic veins of the superficial system with the humeral veins of the deep system. It also receives blood from the shoulder, scapular, mammary areas, etc.

posterior circumflex humeral vein

A venous branch that terminates in the external part of the cephalic vein. It collects blood from the deltoid muscle and the shoulder joint.

anterior circumflex humeral vein

An external venous branch that terminates in the cephalic vein and collects the venous blood from part of the area of the deltoid muscle.

cephalic vein

The cephalic vein originates in the union of the median cephalic vein and the superficial radial vein. It ascends the arm in its external superficial area and terminates in the axillary vein, receiving venous branches from the arm and the elbow.

median cephalic vein

One of the veins which ascends the anterior face of the flexure of the elbow, from the bifurcation of the median vein to unite with the superficial radial vein to form the cephalic vein.

superficial radial vein

The superficial radial crosses the forearm superficially, firstly posteriorly and then externally. It collects the venous blood from the forearm and also from the external dorsal part of the hand. It unites with the median cephalic vein to form the cephalic vein.

VENOUS SYSTEM OF THE SUPERIOR EXTREMITY

The superior and inferior extremities are equipped with twin venous systems, one deep and the other superficial. The deep venous system is a system parallel to the arterial system, with identical names and trajectory, although there are two veins for each artery. For these reasons, it is not considered necessary to illustrate it. The superficial venous system, on the contrary, has a different trajectory and nomenclature and runs up the arm in more superficial areas, although it terminates in the deep venous system.

thoracic veins

A group of veins that ascend from the lateral thoracic region and terminate in the axillary vein.

basilic vein

Formed by the union of the median basilic vein with the superficial ulnar vein in the flexure of the elbow. It then ascends the internal part of the arm to reach the final part of the humeral veins of the deep venous system near to where these terminate in the axillary vein.

anastomosis of the cephalic and basilic veins

The basilic and cephalic veins are united by a transversal communicating vein or anastomosis, that allows venous blood to be exchanged and ensures a correct venous return.

median basilic vein

A bifurcation of the median vein, that extends to the internal part of the flexure of the elbow and terminates, together with the ulnar vein, in the basilic vein.

median vein

Vein that ascends the anterior face of the forearm, from the palmar area of the hand to the elbow, where it bifurcates into the median cephalic and median basilic veins

superficial ulnar vein

A vein that collects the blood from the internal, dorsal part of the hand, ascends the internal portion of the forearm and terminates, together with the median basilic vein, in the basilic vein.

87

VENOUS SYSTEM: CRANIAL SINUS

superior longitudinal or saggital sinus

A duct that crosses the inside of the cranial vault, from front to back, following a channel excavated in the vault. It collects the venous blood from the orbital area and the internal face of the cerebral hemispheres and terminates in the lateral sinus.

CRANIAL SINUSES

Venous ducts that cross the skull internally through the space adjacent to the dura mater, collecting the venous blood from the brain and the other intracranial organs and carrying it to the internal jugular vein, where they all converge.

lateral sinus

The lateral or transverse sinus originates at the union of the superior longitudinal sinus and the straight sinus and, after bordering the sides the occipital fossa, terminate in the origin of the internal jugular vein.

internal jugular vein

The venous blood collected by the cranial sinus is deposited in the internal jugular vein, leaves the skull through a torn hole in the posterior and continues down the neck.

inferior petrous sinus

The inferior petrous sinuses originate in the cavernous sinuses and terminate in the origin of the internal jugular vein. They cross the inferior part of the petrous bone.

superior petrous sinus

The superior petrous sinuses communicate the cavernous sinuses with the lateral sinuses and collect the venous blood from the base of the cerebral hemispheres. They cross the superior part of the petrous bone.

straight sinus

The straight sinus collects the venous blood from the base of the brain and part of the cerebellum. It terminates in the union of the superior longitudinal sinus with the lateral sinus.

posterior occipital sinus

Sinuses which cross the edges of the occipital foramen laterally to reach the termination of the lateral sinuses in the internal jugular vein. They collect the venous blood from the posterior part of the cerebellum.

coronary or intracavernous sinus

An elliptical sinus located within the sella turcica, surrounding the hypophysis gland. It terminates laterally in the coronary sinus.

sphenoparietal sinus

The sphenoparietal sinuses cross the sphenoid bone by the posterior edge of the roof of the orbit and terminate in the cavernous sinuses, collecting the venous blood from the anterior area of the brain.

cavernous sinus

The cavernous sinuses are located at each side of the sella turcica where the hypophysis is situated and collect the blood from the ophthalmic vein, which comes from the orbit, the coronary sinus, and the area of the sphenoid bone. They continue through the petrous sinus.

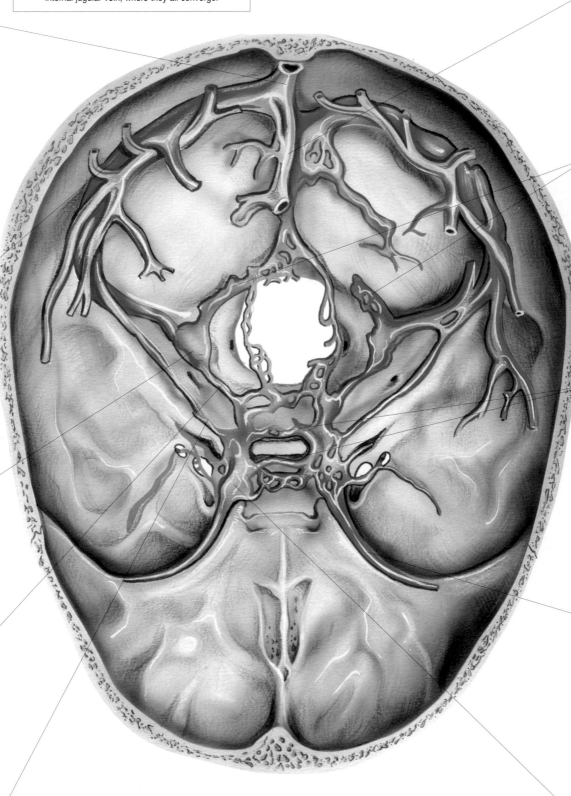

VENOUS SYSTEM: NECK AND HEAD

parietal veins
A network of small vessels that collect the venous blood from the parietal area and carry it to the superficial temporal vein.

internal maxillary vein
The internal maxillary collects the venous blood from the maxillary region and unites with the superficial temporal vein, with which it forms a common trunk that terminates in the external jugular vein, also joining with the internal jugular vein, to communicate both venous systems.

auricular and occipital veins
They collect the venous blood from the auricular and occipital areas and terminate in the internal jugular vein.

anastomosis between the systems of the external and internal jugular veins
The two main venous systems of the skull communicate by small communicating veins which ensure a correct venous return.

external jugular vein
Vein that crosses the external superficial part of the neck and terminates at the union of the subclavian and internal jugular veins. It originates at the convergence of the veins from the occipital, temporal and maxillary areas from the internal jugular system.

vertebral vein
Vein that descends the neck in parallel with the vertebral column and collects the venous blood from that area, terminating, together with the external jugular vein, in the subclavian vein.

brachiocephalic venous trunk
A common trunk formed by the union of the veins of half of the head (internal jugular) and of the superior extremity (subclavian vein). It terminates in the superior vena cava, which carries the venous blood from both areas to the right atrium of the heart.

frontal veins
A series of small blood vessels that collect the venous blood from the frontal area and terminate in the facial vein which carries the blood to the internal jugular vein.

superficial temporal vein
A vein that collects venous blood from the parietal veins and the temporal area and joins the internal maxillary vein to form a common trunk that terminates in the external jugular vein and is also united to the system of the internal jugular vein.

facial vein
A vein that originates in the inner angle of the eye and, after crossing the facial area, terminates in the internal jugular vein. It has lingual and thyroid branches.

anterior jugular vein
The anterior jugular originates in the submentonian area and, after collecting the venous blood from the anterior area of the neck, it terminates close to the union between the external jugular vein and the subclavian vein.

internal jugular vein
A vein that collects the venous blood from the intracranial sinuses, which drain all the structures of the skull. It leaves the skull by the posterior torn hole and descends the neck to reach the brachiocephalic trunk, receiving veins from the thyroid, the tongue, the face and the temporal area and maxillary areas in its path.

subclavian vein
A vein that comes from the arm and terminates at the brachiocephalic venous trunk. It receives branches from the scapular, thyroid and intercostal area, although many branches go directly to the brachiocephalic trunk.

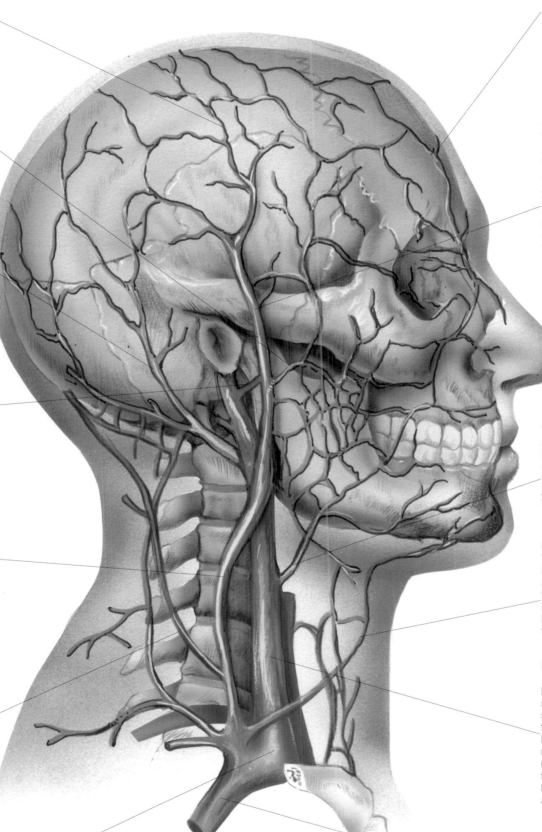

89

VENOUS SYSTEM: ABDOMEN AND PORTAL VEIN

portal vein
The vein which carries the venous blood from the abdominal digestive organs to the liver. It originates in the union of the superior and inferior mesenteric veins and the splenic vein and ascends to the liver, which it enters through the hepatic hilum, dividing into multiple branches inside the liver. In its extrahepatic portion, it receives branches from the stomach, gallbladder, the umbilical area and the pancreas.

inferior vena cava
A common trunk which receives the venous blood from the inferior half of the body. It originates in the inferior area of the abdomen from the union of the two common iliac veins (right and left), which collect the blood from the organs of the pelvic cavity and the inferior extremity.

gastric coronary veins
Veins that cross the lesser curvature of the stomach and terminate directly in the portal vein immediately before hepatic hilum.

splenic vein
The spenic vein originates in the spleen and follows an almost horizontal path to join the inferior and superior mesenteric veins to form the portal vein.

umbilical vein
A vestige of the umbilicus which supplies nutrients to the fetus. After birth, the vein has no function and atrophies.

renal vein
A branch that originates in the renal hilum and carries the venous blood of the kidney to the inferior vena cava which it joins perpendicularly.

superior mesenteric vein
A vein that transports the venous blood from the small intestine and the right half of the large intestine. It joins with the inferior mesenteric and splenic veins to form the portal vein. It has jejunal, ileocolic, colic, pancreatic and epiploic branches.

gastroepiploic vein
A vein that collects blood from the left border of the stomach and from the omentum which support the intestines. It terminates in the superior mesenteric vein.

inferior mesenteric vein
The inferior mesenteric collects the venous blood from the left half of the large intestine. It has hemorrhoidal, sigmoid, rectal and colic branches and joins the splenic vein and the superior mesenteric vein posteriorly to form the portal vein.

right colic vein
The right colic collects the blood from the ascending colon and terminates in the superior mesenteric vein.

common iliac vein
A vein which originates in the union of the internal and external iliac veins and ascends obliquely to join with the opposing common iliac vein, forming the inferior vena cava.

genital vein
A venous branch that ascends from the male (testicular vein) and female (ovarian vein) genital organs to terminate in the renal vein.

external iliac vein
A continuation of the femoral vein which collects all the venous blood from the inferior extremity. In the abdomen, it joins the internal iliac vein to form the common iliac vein.

internal iliac vein
Also known as the hypogastric vein. It collects the venous blood of the intrapelvic organs (bladder, uterus, rectum and anus,) of the gluteal area and the external genitals (pudendal veins). It is linked to the external iliac vein and between the two of them, they give rise to the common iliac vein.

middle colic vein
A vein that runs alongside the transverse colon, collecting its venous blood and transporting it to the superior mesenteric vein.

left colic vein
A branch that terminates in the inferior mesenteric vein after collecting the venous blood from the descending colon.

VENOUS SYSTEM: THORAX, VENA CAVA AND AZYGOS

internal jugular vein
A vein that collects the venous blood from the intracranial sinus, which drains all the structures of the skull. It descends the neck to reach the brachiocephalic trunk. It receives branches from the thyroids, the tongue, the face, the temporal and maxillary areas.

brachiocephalic venous trunk
Two venous trunks whose union gives rise to the superior vena cava. The right trunk receives the venous blood from the right super or extremity and the right half of the head and neck and the left trunk performs a similar function on the opposite side of the body.

vertebral vein
Vein that descends the neck in parallel with the vertebral column and collects the venous blood from that area, terminating together with the external jugular vein in the subclavian vein.

subclavian vein
A continuation of the axillary vein which collects the blood from the superficial and deep venous systems of the superior extremity and joins the internal jugular vein to form the brachiocephalic trunk.

inferior thyroid veins
Veins that collect the venous blood from the inferior part of the thyroid and carry it to the brachiocephalic venous trunk.

internal mammary vein
A vein which ascends the thoracic wall and terminates in the brachiocephalic trunk near its union with the superior vena cava. It collects venous blood from the abdomen, diaphragm and anterior intercostal areas.

superior vena cava
A thick venous trunk which receives all the venous blood of the superior half of the body (trunk, superior extremities and head). It originates in the union of the right and left brachiocephalic trunks and terminates in the right atrium of the heart.

hemiazygos vein
A vein that runs parallel to the azygos vein on the left border of the vertebral column and collects the venous blood from some intercostal veins. It terminates in the left margin of the azygos vein as a side or double branch at around the eighth and ninth vertebra.

intercostal veins
Venous branches that join the azygos vein perpendicularly, after crossing the intercostal spaces and collecting the venous blood from these areas.

azygos vein
Together with the hemiazygos vein, the azygos vein, originating in the thorax, forms a venous system complementary to the vena cava. It collects the venous blood from the mediastinum, the diaphragm, and the intercostal and lumbar areas. It ascends the right side of the vertebral column and terminates in the superior vena cava.

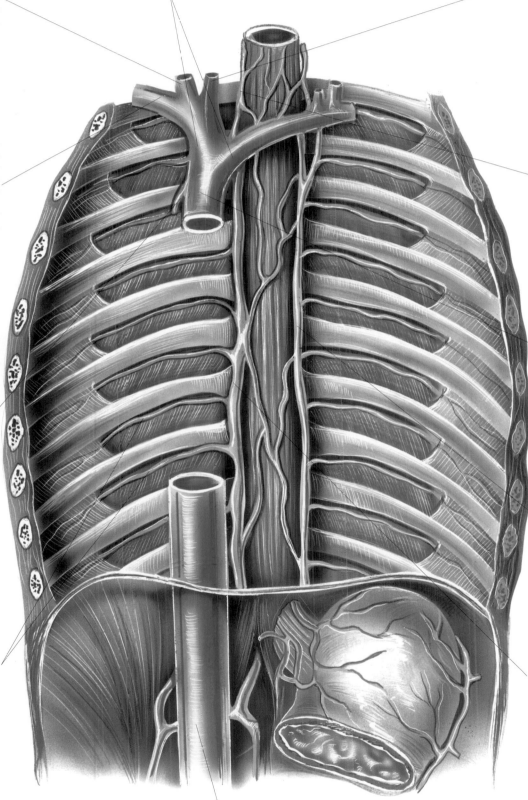

inferior vena cava
A common trunk which receives the venous blood of the inferior half of the body, including the abdomen, pelvis and inferior extremities. After crossing the diaphragm, it enters the thoracic cavity and terminates, together with the superior vena cava, in the right atrium of the heart.

91

DIGESTIVE SYSTEM

▼ GENERAL VIEW

salivary glands
Clusters of glands contained in the walls of the mouth. They secrete saliva needed to masticate and digest food into the mouth by means of small ducts.

oral cavity
The mouth is responsible for the first steps in digesting the food we eat, including the processes of salivation, mastication and swallowing. The inside of the mouth is lined with a fine layer of mucosa called the buccal mucosa which extends to the pharynx.

teeth
Bony structures inside the mouth arranged in superior and inferior rows in the gingival or gums. Their function is to tear and masticate food before it is swallowed.

esophagus
A cylindrical duct that extends from the pharynx to the stomach. It descends through the thoracic cavity and diaphragm and has a short section in the abdomen. The walls are formed of muscles which, when contracted, push the food downwards.

pancreas
A glandular organ located behind stomach. Through a small duct called the duct of Wirsung, the pancreas sends secretions, which contain enzymes aiding the digestion of food to the duodenum.

liver
A large organ located in the right superior angle of the abdomen, in the zone called the right hypochondrium. Its main digestive function is the production of bile, a fluid that is sent to the duodenum through the bile ducts and which is necessary to the digestion of dietary fats.

pylorus
The pylorus is a small, round opening located between the stomach and the duodenum. The surrounding area is called the pyloric region.

gallbladder
A saccular organ contained with the biliary system which stores the bile produced by the liver until it is sent to the duodenum.

small intestine
A long tube that leaves the stomach and coils inside the abdominal cavity in multiple angles or intestinal loops. Its function is the digestion, absorption and transport of foods. For better absorption, the internal surfaces are covered with millions of small projections called villi. It consists of three parts, the duodenum, jejunum and ileum.

duodenum
The first section of the small intestine which receives the secretions from the liver and the pancreas.

jejunum
The second section of the small intestine.

ileum
The third and final section of the small intestine.

iliocecal valve
A valve that communicates the final part of the small intestine, the ileum and the initial part of the large intestine, the cecum.

tongue
A flat appendix inside the oral cavity. The front end is free and the back is attached to the anterior zone of the pharynx. It is formed of various muscles which are responsible for a wide range of movements involved in swallowing and phonation.

isthmus of the fauces
The narrow passage from the mouth to the pharynx bordered by the soft palate, the base of the tongue and the lateral pillars, where the palatine tonsils are located.

pharynx
A tube of muscle and membrane that begins in the nasal fossas, descends through the neck and terminates in the esophagus. It has both respiratory and digestive functions.

cardia
The orifice which communicates the stomach with the esophagus.

stomach
The stomach is a large saccular organ that receives food from the esophagus and stores it during the digestive process. The walls of the stomach contain glands which secrete gastric juices that help break down the food. The contraction of the stomach walls mixes the food within to aid digestion.

ascending colon
The section of the large intestine that ascends the right side of the abdomen vertically from the cecum to the hepatic region.

transverse colon
The section of the large intestine that crosses the abdomen transversally from the hepatic area to the splenic area.

descending colon
The section of the large intestine that descends the left side of the abdomen vertically to reach the rectum.

cecum
The initial section of the large intestine, which is formed by a large sac that receives waste products through the ileocecal valve.

vermiform appendix
A lymphatic organ attached to the cecum. Its inflammation causes the condition known as appendicitis.

rectum
The final part of the large intestine, which is a continuation of the descending colon when it enters the pelvic cavity. It terminates in the rectal ampolla, where the formed feces are stored until their expulsion.

large intestine
The large intestine is a continuation of the small intestine and has a greater diameter. It absorbs water, leaving the unabsorbed remains of food which progressively form feces. Its various sections surround the small intestine.

anus
A structure composed of sphincters which forms the final part of the digestive system by means of a muscular system comprising two sphincters (internal and external) which can be opened or be closed voluntarily. The anus can thus expel the feces.

ORAL CAVITY

▼ LATERAL VIEW

vestibule of the mouth
The space between the upper and lower lips and the gingivae.

maxilla
The bone that separates the nasal fossas and the oral cavity, forming part of the hard palate.

buccal mucosa
A fine, rosy membrane which lines all the oral cavity, including the cheeks, the gingivae, the floor of the mouth and the posterior face of the lips. In the tongue, it is called lingual mucosa.

hard palate
The anterior part of the roof of the mouth, supported by the maxilla.

soft palate
The posterior part of the roof of the mouth. Its structure of muscle and membrane has no bony support.

nasal pharynx
The superior part of the pharynx which communicates with the nasal fossas.

lingual tonsils
Two lymph organs similar to the palatine tonsils, but located behind the tongue.

palatine tonsils
Two round organs located between the anterior and posterior pillars of the veil of the palate. They are lymph organs which form part of the body's defense system.

oral pharynx
The middle part of the pharynx is a tube of muscle and membrane that originates in the nasal fossas and enters the neck. It has respiratory (passage of air) and digestive (passage of food) functions.

93

teeth
Bony structures inside the mouth arranged in upper and lower rows in the gingivae or gums. Their function is to tear and masticate food before it is swallowed.

tongue
A flat appendix inside the oral cavity. The front end is free and the back is attached to the anterior zone of the pharynx. It is formed of various muscles which confer a wide range of movements used in swallowing and phonation.

laryngeal pharynx
An inferior continuation of the oral pharynx which performs the same functions. It constitutes the final part of the pharynx and finishes in a double duct; posteriorly, it connects with the esophagus and anteriorly with the larynx.

mandible
A facial bone that surrounds the oral cavity anteriorly and laterally. Its articulation with the skull is movable, allowing a series of movements that aid mastication and phonation. Various tongue muscles are inserted in the mandible.

hyoid bone
A thin, U-shaped bone in which the muscles of the tongue and pharynx are inserted.

epiglottis
A flap of cartilage lying behind the tongue that acts as a cover over the orifice which opens to allow air to pass to the respiratory tract. The epiglottis opens to permit the passage of air and closes when food is being swallowed.

larynx
A tubular formation consisting of cartilaginous structures. It contains membranous folds that form the vocal cords, allowing phonation by vibrating the passing air.

esophagus
A cylindrical duct that extends from the pharynx to the stomach. The walls are formed of muscles which, when contracted, push the food downwards.

THE STRUCTURE OF A TOOTH

root canal
A duct located inside the dental root which carries the blood vessels and nerves to the dental pulp.

crown
The external and visible section of the tooth which emerges from the gum.

neck
The intermediate zone of the tooth located between the crown and the root.

root
The section of the tooth implanted within the gingiva in a cavity called the dental alveolus.

cement
A layer that covers the root of the tooth externally and is equivalent to the enamel of the crown. It is a hard, yellowish substance whose external face corresponds to the periodontal ligament uniting the root with the alveolar bone.

dental alveoli
The dental alveoli are the sockets in the maxilla and mandible in which each of the teeth are implanted.

enamel
The external layer of the crown of the tooth. Whitish in color, the enamel is extremely hard.

dentin
The middle part of the tooth, located between the enamel and the dental pulp. It forms the main structure of the tooth and is very hard.

gingiva
The part of the buccal mucosa that covers the maxilla and mandible where the teeth are implanted.

pulp
The central area of the tooth. It is formed by connective tissue which contains the vascular and nervous terminations.

mandible
The bone in which the lower teeth are implanted.

94

TYPES OF TEETH

incisors
Located in the front part of the gingiva, the incisors have one root and a flat crown. There are four upper and four lower incisors which bite and tear foods.

canines
The canine teeth are located behind the incisors. They have a sole root and a conical, pointed crown. Two upper and two lower canines tear and bite food.

premolars
Located behind the canines, the premolars have a single root and a cubical crown. Four upper and four lower premolars crush and grind food.

molars
The molars are located at the back of the mouth behind the premolars. They have multiple roots and an irregularly shaped crown. People have as many as six upper and six lower molars, although some people lack the last two upper and lower molars on each side, the wisdom teeth. Their function is to crush and to grind food.

TEETHING

1ST TEETHING

Deciduous dentition. Babies are normally born without teeth, which begin to appear after about six months. The first set of teeth, the deciduous dentition, is temporary and will fall out spontaneously as the child grows, to be replaced by the permanent teeth. There are usually 20 deciduous teeth.

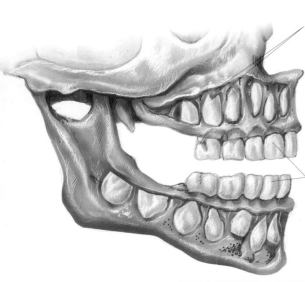

alveoli
Cavities inside the mandible and maxilla which contain the dental buds until their cutting.

deciduous dentition
Temporary teething begins at around six months and is complete by twenty-four to thirty months. The deciduous teeth begin to fall out spontaneously after about six years of age.

2ND TEETHING

Permanent teething or second teething. It begins to cut around six years of age when the deciduous dentition begins to fall out, although it is often not completed until adulthood. Normally, adults have 32 permanent upper and lower teeth

APPROXIMATE AGE OF CUTTING

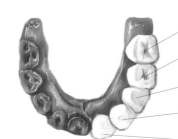

upper central incisor 8-10 months
upper lateral incisor 9-12 months
upper canine 18-24 months
first upper premolar 13-15 months
second upper premolar 24-30 months

second lower premolar 24-30 months
first lower premolar 13-15 months
lower canine 18-24 months
lower lateral incisor 9-12 months
lower central incisor 6-9 months

APPROXIMATE AGE OF ERUPTION

upper central incisor 6-8 years
upper lateral incisor 6-8 years
upper canine 11-12 years
first upper premolar 10-11 years
second upper premolar 12-13 years
first upper molar 6-7 years
second upper molar 12-14 years
third upper molar 18-30 years
(In some people it may not cut).

third lower molar 18-30 years
(In some people it may not cut).
second lower molar 12-14 years
first lower molar 6-7 years
second lower premolar 10-11 years
first lower premolar 12-13 years
lower canine 11-12 years
lower lateral incisor 8-9 years
lower central incisor 8-9 years

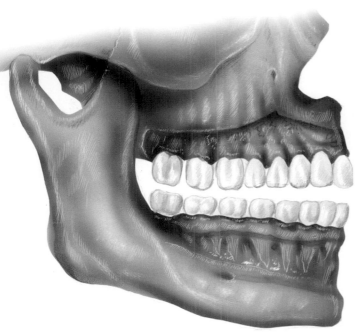

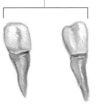
incisors

canine

premolars

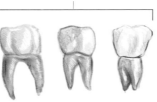
molar

ESOPHAGUS

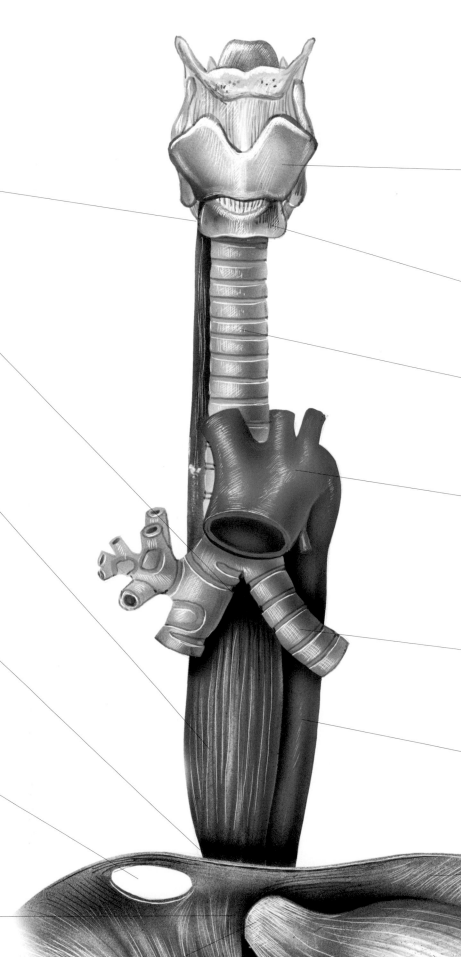

pharyngo-esophageal constriction
The esophagus is an irregularly shaped cylinder with some narrowing where it gives support to other organs. In its superior part, the pharyngo-espophageal constriction represents the point where the esophagus comes into contact with the cricoid cartilage of the trachea.

aortobronchial constriction
A narrowing located in the left border of the middle third of the esophagus, corresponding to the esophagus, the left main bronchus and the aorta.

esophagus
A cylindrical duct that extends from the pharynx to the stomach. It descends through part of the neck, the thoracic cavity and diaphragm and has a short section in the abdomen. The walls are formed of muscles which, when contracted, push the food downwards.

diaphragmatic constriction A narrowing of the esophagus at the level of the diaphragm with enables the esophagus to pass through the esophageal hiatus muscle.

hiatus of the vena cava
An orifice that opens near the esophageal hiatus in the diaphragm. It allows the passage of the inferior vena cava from the abdominal cavity to the thoracic cavity, where it terminates in the right atrium of the heart.

esophageal hiatus
An orifice of the central section of the diaphragm in its central zone which allows the esophagus to pass from the thoracic to the abdominal cavity.

cardia
An orifice which communicates the stomach with the esophagus and acts as a sphincter or valve, opening to admit food and closing to prevent a reflux.

thyroid cartilage
A cartilage which forms the anterior wall of the larynx. Its anterior part marks a protuberance in the neck known as the Adam's apple, which is more pronounced in males.

cricoid cartilage
A cartilaginous ring that forms the inferior limit of the larynx and is supported by the superior part of the esophagus.

trachea
A tubular structure that forms part of the respiratory system and descends, parallel and posterior to the esophagus, to communicate the larynx with the lungs.

arch of the aorta
The aorta ascends from the left ventricle and then immediately curves left and descends, forming an arch located anterior to the superior section of the esophagus.

left main bronchus
One of the two main bronchi into which the trachea bifurcates. The left bronchus is supported by the middle section of the esophagus.

thoracic aorta
Continuation of the arch of the aorta, which descends posteriorly through the diaphragm, paralell to the aorta.

diaphragm
A flat muscle that separates the thoracic and abdominal cavities. It has various orifices or hiatuses that allow organs such as the esophagus and aorta to pass from one cavity to another.

stomach
A large saccular organ. The esophagus passes through the diaphragm and deposits its contents in the stomach after entering the abdomen. The stomach accumulates the ingested food and begins to digest it through the action of the gastric juices secreted by glands located in the walls of the stomach.

STOMACH

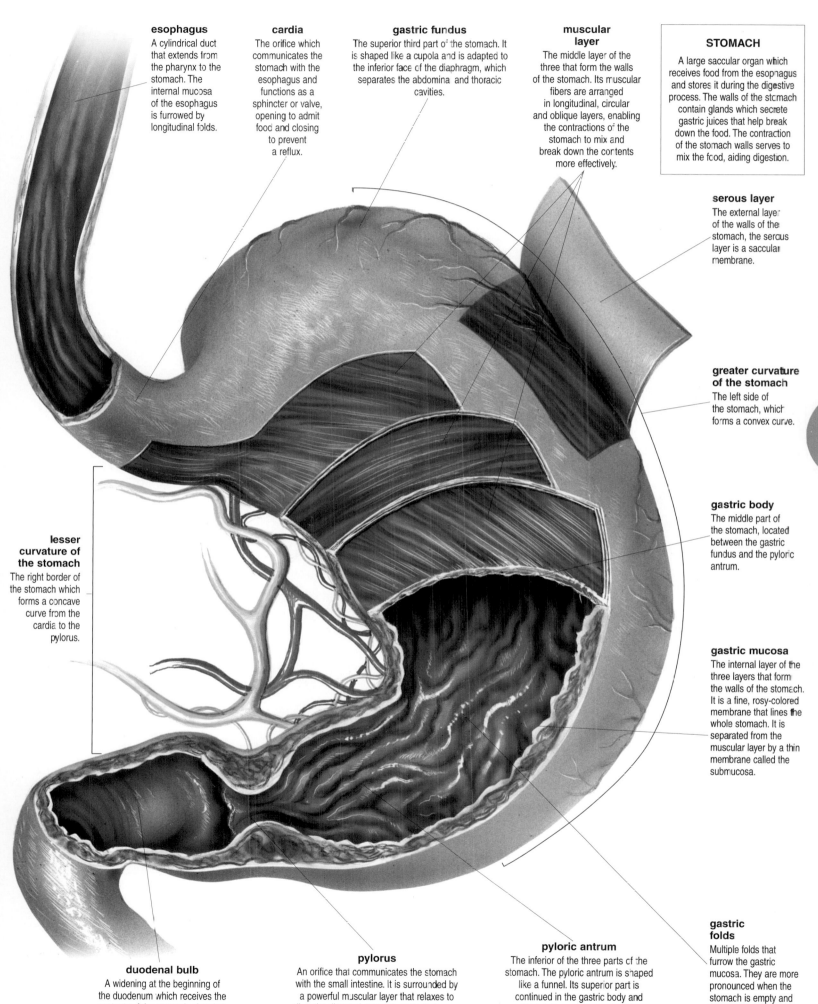

esophagus
A cylindrical duct that extends from the pharynx to the stomach. The internal mucosa of the esophagus is furrowed by longitudinal folds.

cardia
The orifice which communicates the stomach with the esophagus and functions as a sphincter or valve, opening to admit food and closing to prevent a reflux.

gastric fundus
The superior third part of the stomach. It is shaped like a cupola and is adapted to the inferior face of the diaphragm, which separates the abdominal and thoracic cavities.

muscular layer
The middle layer of the three that form the walls of the stomach. Its muscular fibers are arranged in longitudinal, circular and oblique layers, enabling the contractions of the stomach to mix and break down the contents more effectively.

STOMACH
A large saccular organ which receives food from the esophagus and stores it during the digestive process. The walls of the stomach contain glands which secrete gastric juices that help break down the food. The contraction of the stomach walls serves to mix the food, aiding digestion.

serous layer
The external layer of the walls of the stomach, the serous layer is a saccular membrane.

greater curvature of the stomach
The left side of the stomach, which forms a convex curve.

97

gastric body
The middle part of the stomach, located between the gastric fundus and the pyloric antrum.

lesser curvature of the stomach
The right border of the stomach which forms a concave curve from the cardia to the pylorus.

gastric mucosa
The internal layer of the three layers that form the walls of the stomach. It is a fine, rosy-colored membrane that lines the whole stomach. It is separated from the muscular layer by a thin membrane called the submucosa.

duodenal bulb
A widening at the beginning of the duodenum which receives the contents of the stomach after they pass through the pylorus.

pylorus
An orifice that communicates the stomach with the small intestine. It is surrounded by a powerful muscular layer that relaxes to allow the digested stomach contents to pass and then contracts to prevent a reflux.

pyloric antrum
The inferior of the three parts of the stomach. The pyloric antrum is shaped like a funnel. Its superior part is continued in the gastric body and the inferior part connects with the intestines through the pylorus.

gastric folds
Multiple folds that furrow the gastric mucosa. They are more pronounced when the stomach is empty and flatten when it is distended with food.

SMALL AND LARGE INTESTINES

SMALL INTESTINE

A long tube of more than 18 feet in length that coils in multiple angles or intestinal folds inside the abdominal cavity. It continues the process of digesting and absorbing the stomach contents and consists of three parts: the duodenum, jejunum and ileum.

hepatic flexure of the colon

An angle formed by the colon at the height of the liver which serves as the border between the ascending colon and the transverse colon.

transverse colon

The section of the large intestine that crosses the abdomen transversally from the hepatic area to the splenic area and continues as the descending colon.

splenic flexure of the colon

An angle formed by the colon at the height of the spleen which marks the border between the transverse and descending colon.

LARGE INTESTINE

The large intestine is a continuation of the small intestine, which it surrounds like a frame. It absorbs water, leaving the unabsorbed remains of food which progressively form feces. It has various sections; the cecum, ascending colon, transverse colon, descending colon, sigmoid colon and the rectum.

duodenum

The first part of the small intestine, which a forms a large C that surrounds the head of the pancreas. It consists of three parts: the first is oblique and begins in the pylorus, the second is descending and the third is ascending and terminates in the angle of Treitz to give way to the jejunum. The duodenum receives the secretions of the liver and pancreas which aid the digestive process.

jejunum

The second or middle part of the small intestine, which begins at the angle of Treitz. The union with the ileum is not well-defined and there is little to distinguish them. The jejunum carries out most of the food-absorption process.

ileum

The third and final part of the small intestine. It connects with the large intestine through a sphincter muscle known as the ileocecal valve or Bauhin's valve, which allows the contents of the intestine to pass to the large intestine and prevents reflux. The ileum absorbs many nutrients that have not yet been absorbed in the duodenum and the jejunum.

descending colon

The section of the large intestine that vertically descends the left side of the abdomen to reach the rectum.

sigmoid colon

A continuation of the descending colon which enters the pelvic cavity. Its shape varies from one individual to another.

rectum

The final part of the large intestine. In its final part it has an expansion called the rectal ampolla which is where the formed feces are stored until their expulsion.

ascending colon

A duct that ascends the right side of the abdomen vertically from the cecum to reach the hepatic region, where it forms an angle and continues as the transverse colon.

cecum

The initial part of the large intestine, formed by a large sac which receives the contents of the small intestine through the ileocecal valve or Bauhin's valve.

vermiform appendix

A lymphatic organ attached to the cecum. Its inflammation causes the condition known as appendicitis.

anus

A sphincter or valve which is the final part of the digestive tract. It contracts or relaxes voluntarily allowing the expulsion of feces.

LARGE INTESTINE. THE CECUM AND ANAL AREA

taenia of the colon
The colon is bordered by three taenia: the free, omental and mesocolic taenia, all of which serve as insertion membranes of the peritoneum.

haustra
Saccular dilations or semi-lunar crests that correspond to the interior constrictions or folds of the large intestine.

ileum
The third and final part of the small intestine which connects with the large intestine through the ileocecal valve or Bauhin's valve, allowing the contents of the intestine to pass to the large intestine and preventing reflux.

CECUM

99

semilunar folds or crests
Transverse constrictions that cover the interior surface of the large intestine. Externally, they appear as the haustra or semi-lunar crests.

cecum
The initial part of the large intestine, located in the right inferior abdomen, in the right iliac fossa. It is saccular and is also known as the cecal ampolla. It receives and stores the contents of the ileum.

vermiform appendix
A cylindrical lymphatic organ which hangs from the cecum by a small orifice. Its walls are covered by abundant mucosal glands and lymphatic tissue.

ANAL AREA

ileocecal valve
An oval orifice, also called Bauhin's valve, which communicates the ileum (small intestine) with the cecum (large intestine).

rectum
The final part of the large intestine. It has an expansion in its final section called the rectal ampolla where the formed feces are stored until their expulsion.

semilunar valves
Small, crescent moon-shaped folds that surround the anus at the point where it joins the rectum.

hemorrhoidal veins
An abundant network of veins that surround the anal canal. When they undergo sustained pressure, they become enlarged and dilated, causing dilations known as hemorrhoids.

anal folds
Radial folds surrounding the final section of the anal duct. When the anus is contracted they are more pronounced, but may disappear when the anus is open and relaxed during defecation.

internal anal sphincter
A ring of smooth muscular fibers located in the interior of the anal orifice. It opens and closes involuntarily, depending on the amount of feces in the rectum. When the rectum is full, the sphincter opens, prompting the desire to defecate.

external anal sphincter
A ring of striated muscular fibers surrounding the exterior of the anal orifice. It can be contracted or relaxed voluntarily when the individual perceives the desire to defecate, allowing control of the process of defecation.

anus
A tubular structure that communicates the alimentary canal with the exterior to allow the expulsion of the fecal bolus.

STRUCTURE OF THE STOMACH WALLS AND INTESTINES

mucous layer
The mucous layer lines the stomach and is formed by multiple protuberances and depressions. It contains secretory glands that produce digestive enzymes, chlorhydric acid, bicarbonate, water and mucous substances.

submucous layer
A very thin layer rich in nervous and vascular terminations.

muscular layer
A thick layer of muscle comprised of three layers of flat muscle fibers arranged longitudinally, circularly and obliquely. When contracted, they permit the movements that allow the stomach contents to be mixed.

adventitia
The outermost layer of connective tissue that surrounds the sac of the stomach and forms part of the peritoneal sac.

STOMACH

mucous layer
The mucous layer lines the small intestine internally. It is covered by tiny intestinal villi and contains multiple secretory glands which produce digestive enzymes and mucous substances to protect the mucosa from the chlorhydric acid of the stomach.

SMALL INTESTINE

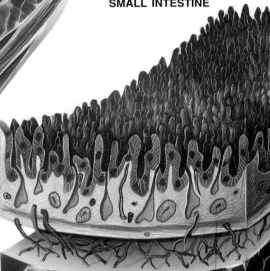

muscular layer
A layer formed of smooth muscle fibers arranged longitudinally and circularly.

submucous layer
A very thin layer rich in nervous and vascular terminations.

adventitia
The outermost layer of connective tissue that surrounds the small intestine and contains blood vessels.

muscular layer
A muscular layer formed of flat muscle fibers arranged longitudinally and circularly. The longitudinal fibers do not surround the intestine, but are grouped together to form the taenia, a thin layer of longitudinal muscle.

LARGE INTESTINE

mucous layer
A mucous layer lines the internal surface of the large intestine. It is much smoother than the corresponding layer in the small intestine and is formed of flat elevations which contain multiple secretory glands that produce mucous substances.

submucous layer
A very thin layer rich in nervous and vascular terminations.

adventitia
The outermost layer of connective tissue that surrounds the large intestine and contains blood vessels.

PERITONEUM

coronary ligament
The largest of the ligaments forming the peritoneal membranes that cover the superior part of the liver and connect it to the diaphragm. The extreme sections of the coronary ligament are known as the right and left triangular ligaments.

falciform ligament
The falciform ligament connects the anterior and superior faces of the liver to the anterior abdominal wall and diaphragm.

spleen
An oval organ located in the left superior angle of the abdomen, in the left hypochondrium. It is posterior to the stomach and connected to it by a peritoneal ligament called the gastrosplenic omentum. It is a lymph organ which stores and replenishes the blood cells.

liver
The liver is attached to the lesser omentum, peritoneal membranes which join to form a series of ligaments attaching the liver to the diaphragm.

gastrosplenic omentum
A membrane that joins the superior part of the greater curvature of the stomach with the hilum, the orifice which gives passage to the blood vessels of the spleen.

stomach
The anterior and posterior surfaces of the stomach are covered by the serous membranes of the peritoneum, which continue upwards to form the lesser omentum that attaches the stomach to the liver, and downwards to form the greater omentum.

gallbladder
A saccular organ forming part of the system of the hepatic excretory or biliary system. It stores and concentrates the bile produced by the liver until it is sent to the duodenum.

gastrohepatic omentum
Also known as the lesser omentum, this membrane attaches the lesser curvature of the stomach to the inferior face of the liver.

gastrocolic omentum
Also known as the greater omentum. A membrane that unites the inferior part of the stomach with the transverse colon. In addition, the membrane of the omentum forms a layer overlying the transversal colon and small intestine.

descending colon
The posterior face of the descending colon is directly attached to the abdominal wall and the rest is covered by the peritoneum. In its final part, a ligament attaches it to the wall of the pelvic cavity.

ascending colon
Also known as the right colon. It is only covered by the peritoneum in its anterior face. The posterior face is attached directly to the abdominal wall, making this part of the colon slightly mobile.

small intestine
The anterior separation of the folds of the jejunum and the ileum reveals the mesentery, which supports the folds and attaches them to the posterior abdominal wall.

101

PERITONEUM
A saccular membrane that surrounds almost entirely in some cases, a large part of the abdominal organs. It consists of two layers: the parietal peritoneum, which is attached to the walls of the abdominal cavity, and the visceral peritoneum, which is introduced between the viscera and surrounds and attaches them. Intraperitoneal organs include, the stomach, the spleen, the small intestine and almost all of the large intestine.

mesentery
Part of the peritoneum that supports the folds of the small intestine, attaching them to the posterior wall of the abdomen. It contains numerous blood vessels and nerves.

transverse colon
The anterior and posterior transverse colon is covered by the peritoneum. The membranes join to form the greater omentum in its superior part. The posterior face contains a membranous partition called the transverse mesocolon which connects it to the peritoneal layer covering the posterior wall of the abdomen.

LIVER

common hepatic duct
A duct which carries bile from the hepatic hilum. After a short extrahepatic section, it joins the cystic duct (coming from the gallbladder) to form the common bile duct.

coronary ligament
A membrane that connects the superior part of the liver with the diaphragm.

hepatic fissure
A large sulcus that divides the liver external and internally into two segments or lobes.

inferior vena cava
A thick, venous trunk that collects the blood from the inferior extremities and the abdominal organs and carries it to the right atrium of the heart. The suprahepatic veins, which carry blood from the liver, join the vena cava posterior to the liver.

LIVER
A large organ located in the right superior angle of the abdomen, in the zone called the right hypochondrium. Its main digestive function is the production of bile, a fluid sent to the duodenum through the biliary ducts and which is fundamental to the digestion of dietary fats. The liver is vital in producing energy for the body because it converts a large part of the glucose and other nutrients absorbed into usable energy.

capsule of Glisson
An external layer composed of fibrous tissue that covers all the liver. It is a red-brown color and has a granular aspect.

intrahepatic ducts
The interior of the liver is furrowed by small ducts that converge in the hepatic hilum and form the common hepatic duct. Their function is to collect and transport the bile secretions.

left hepatic lobe
The liver is divided into two segments or lobes, of which the left or internal lobe is the smaller.

right hepatic lobe
The liver is divided into two segments or lobes, of which the right or external lobe is the largest.

spleen
An oval organ located in the left superior angle of the abdomen, in the left hypochondrium.

cystic duct
A thin duct that leaves the gallbladder and joins with the common hepatic duct to form the common bile duct.

hepatic hilum
An orifice, located the inferior face of the liver, through which the hepatic veins and arteries and the common hepatic duct pass.

gallbladder
A saccular organ within the biliary system which stores and concentrates the bile produced by the liver until it is sent to the duodenum.

abdominal aorta
Part of the aorta that descends the abdomen vertically. It branches into a thick artery called the celiac trunk which, in turn, branches into the hepatic artery.

common bile duct
A duct formed by the union of the cystic and common hepatic ducts. The bile, together with the pancreatic secretions, is emptied into the duodenum through the major duodenal papilla or papilla of Vater.

celiac trunk
A thick, arterial trunk that originates in the abdominal aorta and has branches to the liver, stomach and spleen.

portal vein
A thick, venous trunk that enters the liver through the hepatic hilum. It is located where the superior and inferior mesenteric veins, which collect the venous blood from the large and small intestines, and the splenic vein, which collects blood from the spleen, meet.

stomach
A large saccular organ located in the superior part of the abdomen, under the epigastric region, behind and below the liver.

duodenum
The first part of the small intestine going from the stomach to the jejunum. It receives the secretions of the liver and pancreas which enter through the major duodenal papilla or papilla of Vater.

main pancreatic duct or duct of Wirsung
A duct which carries pancreatic secretions from the pancreas to the duodenum which it enters through the major duodenal papilla jointly with the common bile duct.

pancreas
A glandular organ located below the liver and behind the stomach. It produces the pancreatic juices that go to the duodenum to aid the digestion of foods. In addition, it manufactures a hormone called insulin which goes to the blood and is fundamental in the metabolism of sugars.

hepatic artery
A branch of the celiac trunk that enters the liver through the hepatic hilum, forming branches in its interior. It supplies the liver with arterial blood. (In the drawing, the artery is displaced downwards to show the portal vein and common hepatic duct).

STRUCTURE OF THE LIVER

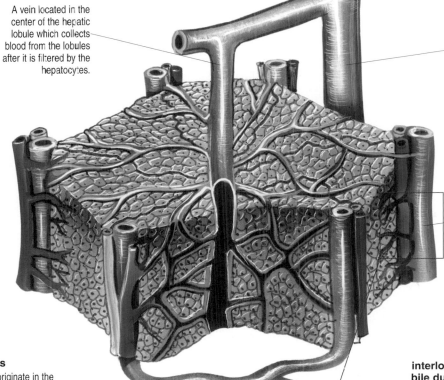

HEPATIC LOBULE

The liver is composed of small hexagonal structures measuring less than 1 mm in diameter. They are called hepatic lobules and are composed of hepatic cells (hepatocytes) grouped around a central vein. These lobes are the functional units of the liver, filtering the blood from the portal vein and manufacturing bile, which is the secretion excreted to the intestine to aid digestion.

central lobular vein

A vein located in the center of the hepatic lobule which collects blood from the lobules after it is filtered by the hepatocytes.

interlobular vein

The sublobular veins are located between the lobules. They collect the blood from the central lobular veins and converge to form the suprahepatic veins which carry the blood from the liver to the inferior vena cava.

periportal space

The space between the hepatic lobules, through which the branches of the portal vein, the hepatic artery and the interlobular bile ducts pass. This space is surrounded by the conjunctive tissue which encloses the hepatic lobules.

sinusoids

Venous capillaries that originate in the branches of the portal vein, radiate out to the hepatic lobule and take the venous blood to the hepatocytes, where it is filtered and the substances needed by the hepatocytes are extracted. The sinusoids terminate in a vein located in the center of the hepatic lobule called the central lobular vein.

interlobular bile ductule

Small ducts located in the periphery of the hepatic lobule, which collect the bile transported by the biliary canaliculi. The interlobular bile ductules converge to form thicker biliary ducts which terminate in the large right and left intrahepatic ducts, that carry the bile out of the liver.

branch of the portal vein

The portal vein carries venous blood to the liver. Inside the liver, it branches into successive ramifications that surround the hepatic lobule and carry the blood to it.

103

Kuppfer's cells

Kuppfer's cells are lymphoid cells contained within the sinusoids which neutralize any foreign bodies that could damage the body, such as bacteria or dead cells.

biliary canaliculus

A thin duct that runs between the hepatocytes and collects the secreted bile, transporting it to the interlobular bile ducts.

hepatocytes

The cells which form the hepatic tissue. They perform the many complex functions of the liver, such as glycogen storage to supply the body's glucose reserve, the manufacture of proteins or the filtration of the blood to produce bile.

branch of the hepatic artery

The hepatic artery carries arterial blood to the liver. Inside the liver, the hepatic artery branches successively to reach the periphery of the hepatic lobule, contributing the blood necessary for the operation of the hepatocytes. This blood is then excreted to the sinusoids where it is mixed with the venous blood.

GALLBLADDER

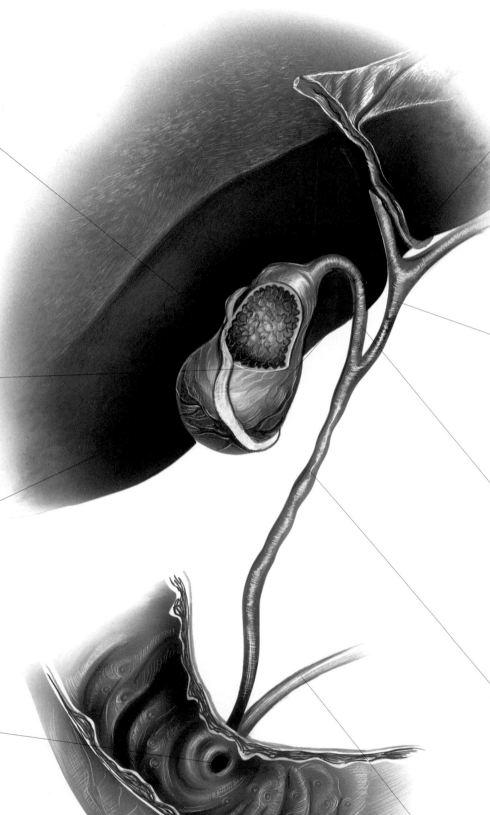

mucous layer
The internal layer (one of three) which form the wall of the gallbladder. It covers the internal surface of the gallbladder and is furrowed by numerous folds. It contains glands which produce mucous substances.

muscular layer
The central of the three layers forming the wall of the gallbladder. It is composed of muscular fibers which, when contracted, cause the bile accumulated in the gallbladder to be expelled through the cystic duct.

serous layer
The external of the three layers of the gallbladder wall. It is formed by the prolongation of the peritoneal membranes that cover the liver.

major duodenal papilla or **papilla of Vater**
An orifice which marks the termination of the short common canal of the extrahepatic biliary tree and the pancreatic ducts or ducts of Wirsung, through which they excrete their secretions into the duodenum. The ducts unite to form a small widening called the ampulla of Vater.

duodenum
The first part of the small intestine going from the stomach to the jejunum. It receives secretions from the liver and pancreas which enter through the major duodenal papilla or papilla of Vater.

gallbladder
produced by the liver until it is sent to the duodenum. It is attached to the inferior face of the liver and consists of a part near the cystic duct, called the neck of the gallbladder, a central part called the body and a distal part called the floor of the gallbladder.

common hepatic duct
A duct which carries bile from the hepatic hilum and which results from the union of the right and left intrahepatic ducts. After a short extrahepatic section, it joins the cystic duct coming from the gallbladder to form the common bile duct.

cystic duct
A thin duct that leaves the gallbladder and joins the common hepatic duct to form the common bile duct. Bile is transported through the cystic duct to the gallbladder, where it is stored and concentrated, before its expulsion. The valves it contains may give it a rosy aspect.

common bile duct
A duct formed by the union of the cystic and common hepatic ducts. The bile, together with the pancreatic secretions, is emptied into the duodenum through the major duodenal papilla or papilla of Vater. Its termination, like that of the main pancreatic duct, is a small muscular spincter which only opens when the pancreatic fluids are needed.

main pancreatic duct or **duct of Wirsung**
A duct which carries the pancreatic secretions from the pancreas to the duodenum which it enters jointly with the common bile duct through the major duodenal papilla. Its termination, like that of the common bile duct, is a small muscular spincter which only opens when the pancreatic fluids are needed.

PANCREAS

gallbladder
A saccular organ contained with the biliary system which stores and concentrates the bile produced by the liver until it is sent to the duodenum.

cystic duct
A thin duct that leaves the gallbladder and joins with the common hepatic duct to form the common bile duct.

common bile duct
A duct formed by the junction of the cystic and common hepatic ducts. The bile, together with the pancreatic secretions, is emptied into the duodenum through the major duodenal papilla or papilla of Vater.

common hepatic duct
A duct which carries the bile from the liver. After a short extrahepatic passage, it joins the cystic duct coming from the gallbladder to form the common bile duct.

main pancreatic duct or of Wirsung
A duct which carries the pancreatic secretions produced by the secretory sacs called acini from the pancreas to the duodenum which it enters through the major duodenal papilla or papilla of Vater.

celiac trunk
A thick, arterial trunk that originates in the abdominal aorta and has branches to the liver, stomach and spleen, which, in turn, branch into the pancreatic arteries.

PANCREAS
A glandular organ with various functions. The pancreas manufactures the pancreatic juices which go to the duodenum to aid the digestion of food. It also manufactures a hormone called insulin that is sent to the blood and is essential for the metabolism of sugars. The digestive function of the pancreas is carried out by multiple secretory sacs called acini. The pancreas is composed of three sections: the head, body and tail.

105

tail of the pancreas
The thinner, superior end of the pancreas has a flat, slightly pointed shape.

second portion of the duodenum
The second of the three portions of the duodenum which descends vertically. It contains the papillae where the main bile duct from the liver and the main and accessory pancreatic ducts terminate.

smaller papilla of the duodenum
A small eminence in whose vertex a small orifice opens to allow the accessory pancreatic duct (or duct of Santorini) to reach the duodenum.

accessory pancreatic duct or duct of Santorini
A small duct that originates in the main pancreatic duct and terminates in the duodenum through the smaller papilla, excreting pancreatic secretions into the small intestine.

body of the pancreas
The central part of the pancreas which extends from the head, to which it is joined by the narrower neck of the pancreas, to the tail.

third portion of the duodenum
The last of the three sections of the duodenum follows a slightly ascending horizontal path to terminate in the flexure called the angle of Treitz, which marks the beginning of the jejunum.

superior mesenteric artery
A branch of the abdominal aorta that passes behind the pancreas. Its branches irrigate part of the pancreas, the small intestine and part of the large intestine.

major duodenal papilla or papilla of Vater
An orifice which terminates of the short common canal of the extrahepatic biliary tree and the pancreatic ducts or ducts of Wirsung and through which they excrete their secretions into the duodenum. The ducts unite to form a small widening called the ampulla of Vater.

head of the pancreas
The most voluminous part of the pancreas, located between the three sections of the duodenum, which contains the two ducts that end in the duodenum.

superior mesenteric vein
A thick vein that collects venous blood from the small intestine, part of the large intestine and also from the pancreas when it passes behind.

THE RESPIRATORY SYSTEM

▼ GENERAL VIEW

nasal fossas
The cavities in the initial part of the respiratory system through which the air is inspired and expired. As it passes through the nasal fossas, the air is filtered and warmed to aid correct respiration.

nasal pharynx
The nasal fossas end at the nasal pharynx which is the upper part of a larger duct called the pharynx, which has both digestive and respiratory functions, although they are mainly respiratory in the nasal pharynx.

oral pharynx
The central portion of the pharynx, located immediately behind the oral cavity. It has both digestive and respiratory functions, ingesting food and inspiring and expiring air.

laryngeal pharynx
The lower part of the pharynx, which communicates directly with the prolongation of the digestive system (esophagus) and with the inferior respiratory tract (larynx), sharing functions with both systems.

carina
The area where the trachea bifurcates into the two main bronchi.

pulmonary hila
The lungs have internal openings called pulmonary hila through which the pulmonary bronchi and blood vessels enter.

superior lobe of the right lung
The superior lobe occupies the superior half of the right lung.

middle lobe of the right lung
The middle lobe is located in the middle and anterointernal area of the right lung.

inferior lobe of the right lung
The inferior lobe is located in the most inferior and external portion of the right lung.

horizontal fissure
A sulcus that separates the superior lobe from the middle lobe.

oblique fissure of the right lung
A sulcus that separates the middle lobe from the inferior lobe of the lung.

epiglottis
A cartilaginous structure that acts as a cover for the orifice, giving passage to the inferior respiratory tract. When open, it allows air to enter or leave. When shut, it prevents food from entering the respiratory system. The epiglottis is located at the posterior base of the tongue and its movements are controlled by powerful muscles.

glottis
Also known as the vestibule of the larynx, the glottis is located immediately under the epiglottis and gives entry to the larynx.

larynx
A tubular duct formed of cartilaginous structures which contain membranous folds known as the vocal cords, which vibrate as the air passes through them, permitting phonation.

trachea
The inferior prolongation of the larynx, with the same tubular shape. It is formed of a series of cartilaginous rings. It filters inspired and expired air through a mucous layer containing prolongations or cilia and secretory mucus glands.

superior lobe of the left lung
The superoanterior portion of the left lung.

inferior lobe of the left lung
The inferoposterior portion of the left lung.

right and left main bronchi
The final part of the trachea bifurcates to form two lateral ducts called the right and left main bronchi which have the same tubular form and cartilaginous structure. After a brief extrapulmonary section, the two bronchi enter the lungs.

oblique fissure of the left lung
A fissure that separates the superior and inferior lobes of the left lung.

lungs
Two organs composed of spongy, connective tissue located on either side of the thoracic cavity, supported by the diaphragm. They contain the bronchi, bronchioles, alveoli and blood vessels involved in respiration. The lungs are covered by a layer known as the pleura.

THE UPPER AIRWAY

▼ LATERAL VIEW

conchae
Bony protuberances covered by nasal mucosa and located in the lateral walls of the nasal fossas. They create turbulence in the inspired air, warming and humidifying it before it reaches the pharynx.

paranasal sinus
Cavities located in the frontal, sphenoid and ethnoid bones, which join the nasal cavity through small orifices called ostia. They warm the air before it reaches the inferior respiratory tract.

choanae
Two large orifices that form the posterior limit of the nasal fossas. They are bordered by the nasal septum, the roof of the mouth and the lateral walls of the nasal fossas and communicate directly with the oral pharynx.

orifices of the Eustachian tube
Two orifices located laterally in the superior or nasal pharynx which communicate with the cavities of the middle ear through ducts, allowing air to pass through them.

nasal vestibule
Widenings that constitute the initial part of the two nasal fossas. The vestibule, like the other parts of the upper airway is covered by a mucosa rich in mucous glands and villi that filter the air.

nasal pharynx
The nasal fossas end at the nasal pharynx which is the upper part of a larger duct called the pharynx, which has both digestive and respiratory functions, although they are mainly respiratory in the nasal pharynx.

oral pharynx
The central portion of the pharynx which is located immediately behind the oral cavity. It has both digestive and respiratory functions, ingesting food and inspiring and expiring air.

nasal orifices
The anterior openings of the nasal fossas which communicate with the exterior and are located in the inferior part of the nose.

laryngeal pharynx
The lower part of the pharynx, which communicates directly with the prolongation of the digestive system (esophagus) and with the inferior respiratory tract (larynx), sharing functions with both systems.

107

▼ FRONTAL VIEW

hard palate
The anterior part of the palate, which is supported by the maxilla and is also known as the bony palate.

soft palate
The posterior part of the palate formed of muscle and ligaments. It has no bony support.

adenoid tonsils
Spongy structures located in the posterior wall of the oral pharynx. They are formed of lymphoid tissue and form part of the body's defense system, filtering the microscopic impurities and organisms from the inspired air.

lingual tonsils
Structures located in the laryngeal pharynx which are similar to the palatine tonsils.

palatine tonsils
Two structures whose shape, constitution and function are very similar to those of the adenoid tonsils and which are located in the lateral walls of the oral pharynx.

frontal paranasal sinuses
Two cavities located in the frontal bone which end in the nasal cavity through orifices located behind the conchae called meatus.

palate or roof of the mouth
A horizontal partition that separates the nasal fossas from the oral cavity. The hard palate is supported by bone, while the soft palate is composed of muscles and ligaments.

nasal septum
A cartilaginous partition covered by a membranous layer or nasal mucous that separates the nasal fossas into two cavities.

maxillary sinus
Two cavities located in the zygomatic bones which give passage to the nasal cavity through orifices situated behind the conchae called the meatus.

conchae
Bony protuberances covered by nasal mucosa and located in the lateral walls of the nasal fossas. There are generally three: the superior, middle and inferior.

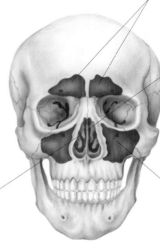

LARYNX AND TRACHEA

▼ ANTERIOR VIEW

▼ INNER VIEW

hyoid bone
A U-shaped bone located in the anterior part of the neck, which serves as the insertion point of the muscles of the tongue, pharynx and larynx.

epiglottis
A cartilaginous structure that acts as a cover for the orifice of the inferior respiratory tract. When open, it allows air to enter or leave. When shut, it prevents food from entering the respiratory system.

glottis
The glottis is the cavity giving entry to the larynx. It is formed of a superior part located over the vocal chords known as the vestibule of the larynx and an inferior portion called the subglottic space.

thyroid cartilage
A cartilaginous structure formed by two lateral laminae joined anteriorly to form the anterior and lateral walls of the larynx. Its anterior border forms a protuberance in the neck, called the Adam's apple, which is more prominent in men.

larynx
A tubular duct located in the anterior part of the neck, formed of cartilaginous structures united by ligaments and muscles. It communicates the pharynx and epiglottis with the trachea. The larynx contains a cavity called the glottis and membranous folds known as the vocal cords.

Morgagni's ventricle
A recess in the lateral wall of the larynx into which the laryngeal sacculus opens.

thyroid cartilage
Two laminae that form the anterior and lateral walls anterior of the larynx, but not its posterior part.

cricoid cartilage
A ring-shaped cartilage which surrounds the larynx and is located below the thyroid cartilage.

arytenoid cartilages
Two small, triangular cartilages at the back of the larynx to which the vocal folds are attached.

vocal cords
Membranous folds located in the middle zone of the glottis. The upper and lower vocal chords vibrate as air passes producing sounds which are modulated by the brain to form the words that we speak. Stretching the vocal chords alters the pitch of the voice.

trachea
The tubular inferior prolongation of the larynx. It is formed of a series of cartilaginous rings. It allows the passage of the inspired and expired air.

carina
The area in which the trachea bifurcates into the two main bronchi.

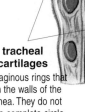

tracheal cartilages
Cartilaginous rings that form the walls of the trachea. They do not form a complete circle, remaining open in the posterior part, where muscular fibers complete the circle, allowing the cartilage to contract or dilate.

tracheal mucosa
A fine mucosa that lines the interior of the trachea. It is rich in glands and covered by fine filaments, or cilia, that filter the air.

cricoid cartilage
A ring-shaped cartilage that surrounds the larynx and marks the inferior limit between the larynx and trachea.

right and left main bronchi
The final part of the trachea bifurcates into two lateral ducts called the right and left main bronchi which have the same tubular form and cartilaginous structure. After a brief extrapulmonary section, the two bronchi enter the lungs.

LUNGS

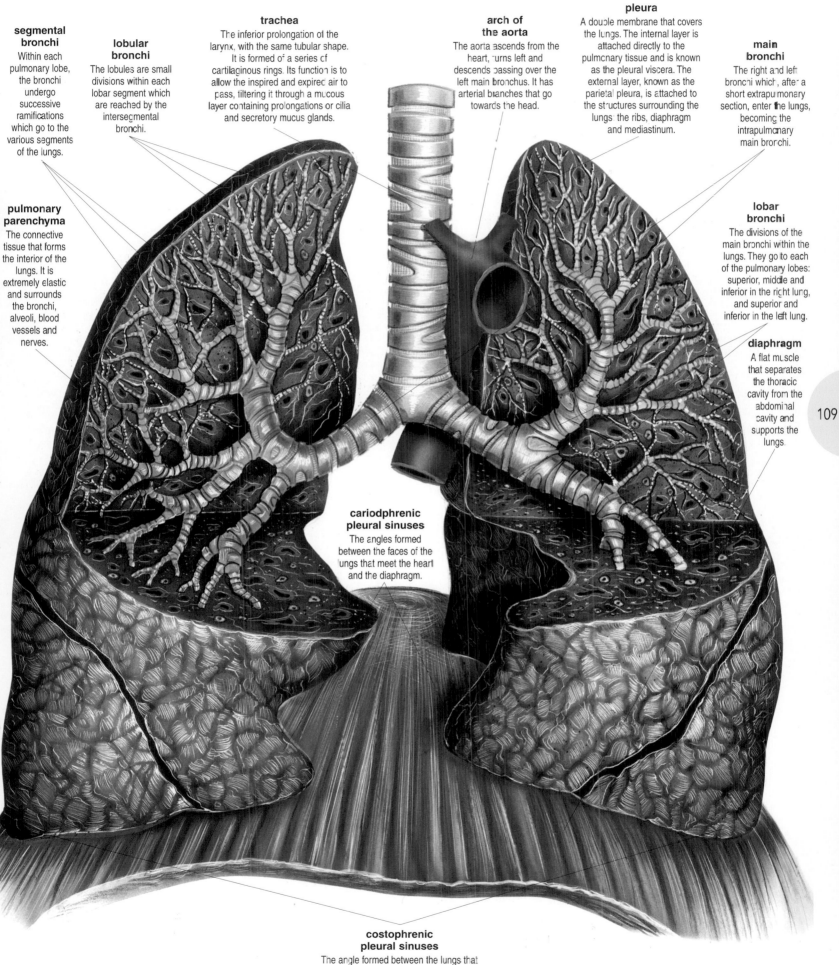

segmental bronchi
Within each pulmonary lobe, the bronchi undergo successive ramifications which go to the various segments of the lungs.

lobular bronchi
The lobules are small divisions within each lobar segment which are reached by the intersegmental bronchi.

trachea
The inferior prolongation of the larynx, with the same tubular shape. It is formed of a series of cartilaginous rings. Its function is to allow the inspired and expired air to pass, filtering it through a mucous layer containing prolongations or cilia and secretory mucus glands.

arch of the aorta
The aorta ascends from the heart, turns left and descends passing over the left main bronchus. It has arterial branches that go towards the head.

pleura
A double membrane that covers the lungs. The internal layer is attached directly to the pulmonary tissue and is known as the pleural viscera. The external layer, known as the parietal pleura, is attached to the structures surrounding the lungs: the ribs, diaphragm and mediastinum.

main bronchi
The right and left bronchi which, after a short extrapulmonary section, enter the lungs, becoming the intrapulmonary main bronchi.

pulmonary parenchyma
The connective tissue that forms the interior of the lungs. It is extremely elastic and surrounds the bronchi, alveoli, blood vessels and nerves.

lobar bronchi
The divisions of the main bronchi within the lungs. They go to each of the pulmonary lobes: superior, middle and inferior in the right lung, and superior and inferior in the left lung.

diaphragm
A flat muscle that separates the thoracic cavity from the abdominal cavity and supports the lungs.

cariodphrenic pleural sinuses
The angles formed between the faces of the lungs that meet the heart and the diaphragm.

costophrenic pleural sinuses
The angle formed between the lungs that meet the internal face of the ribs and the superior surface of the diaphragm.

109

PULMONARY LOBES AND SEGMENTS

RIGHT LUNG

LEFT LUNG

horizontal fissure of the right lung
A sulcus separating the superior lobe and the middle lobe.

pulmonary hila
Orifices located in the inner faces of both lungs which give entry to the bronchi, the blood vessels and the nerves.

oblique fissure of the left lung
A fissure separating the superior and inferior lobes of the left lung.

main bronchi
The trachea bifurcates into the two main bronchi which enter the lungs.

pulmonary arteries
The pulmonary artery carries blood from the right ventricle of the heart to the lungs to exchange the carbon dioxide for oxygen. It divides into the two right and left pulmonary arteries and is the only artery that transports deoxygenated blood.

pulmonary veins
The two right and two left right pulmonary veins carry the oxygenated blood to the left atrium of the heart, from where it passes to the left ventricle and is distributed throughout the body.

oblique fissure of the right lung
A sulcus that separates the middle lobe from the inferior lobe of the lung.

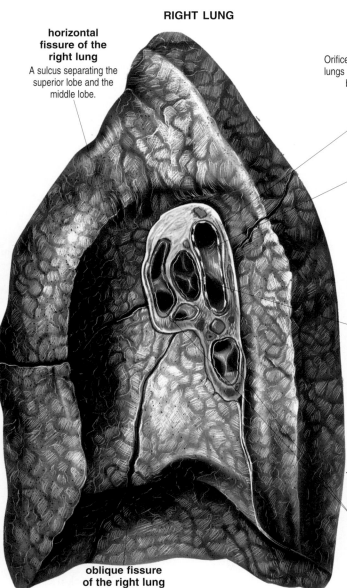
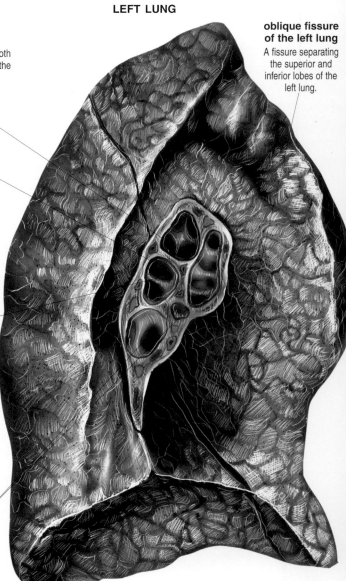

110

BRONCHO-PULMONARY SEGMENTS

RIGHT LUNG

LEFT LUNG

superior lobe
Located in the superior half of the right lung.

apical segment

anterior segment

posterior segment

apical segment

anterior segment

lateral segment

inferior segment

Located in the superoanterior area of the left lung.

middle lobe
Located in the medial anterointernal area of the right lung.

lateral segment

medial segment

inferior lobe
Located in the inferior external area of the right lung.

superior segment

anterior segment

lateral segment

superior segment

anterior segment

lateral segment

Located in the superoanterior portion of the left lung.

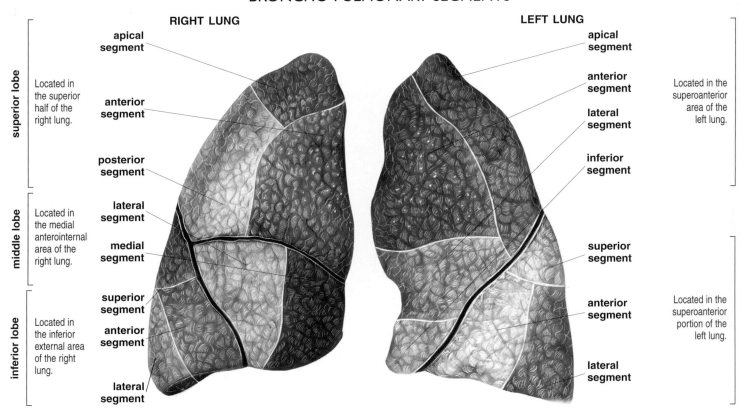

BRANCHES OF BRONCHIAL TREE

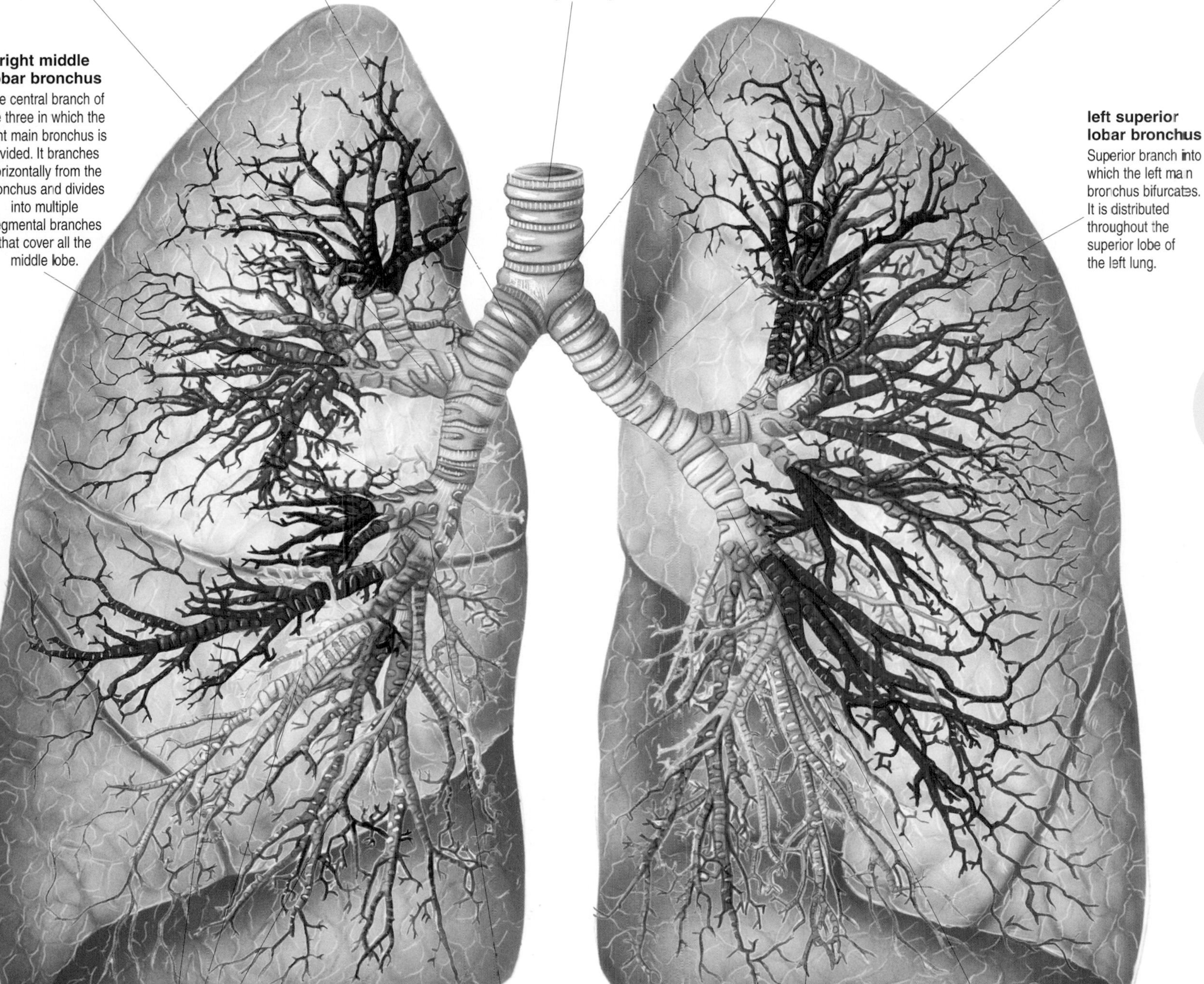

right superior lobar bronchus

The first branch of the right main bronchus ascends to reach the superior lobe of the right lung.

right main bronchus

The trachea divides into the two main bronchi after the carina. The right bronchus, after a short extrapulmonary section, enters the right lung and divides into the superior, middle and inferior lobar bronchi.

trachea

The inferior prolongation of the larynx, with the same tubular shape. It is formed of a series of cartilaginous rings. Its function is to allow the inspired and expired air to pass, filtering it through a mucous layer containing prolongations or cilia and secretory mucus glands.

carina

The area in which the trachea bifurcates into the two main bronchi.

left main bronchus

The final part of the trachea bifurcates to form two lateral ducts, the right and left main bronchi, which have the same tubular form and cartilaginous structure. After a brief extrapulmonary section, the two bronchi enter the lungs.

right middle lobar bronchus

The central branch of the three in which the right main bronchus is divided. It branches horizontally from the bronchus and divides into multiple segmental branches that cover all the middle lobe.

left superior lobar bronchus

Superior branch into which the left main bronchus bifurcates. It is distributed throughout the superior lobe of the left lung.

segmental bronchi

Within each pulmonary lobe, the bronchi undergo successive ramifications which go to the various segments of the lungs.

right inferior lobar bronchus

The last of the three branches of the right main bronchus which follows the same direction. Its branches, the segmental bronchi, spread throughout the inferior lobe of the right lung.

lobular bronchi

The lobules are small divisions within each lobar segment which are reached by the intersegmental bronchi. They are subdivided into bronchioles, which are microscopic terminations that reach the pulmonary alveoli.

left inferior lobar bronchus

Inferior branch of the two into which the left main bronchus bifurcates. Its ramifications are segmental bronchi spread throughout the left inferior lobe of the left lung.

MICROSCOPIC STRUCTURE OF THE LUNGS

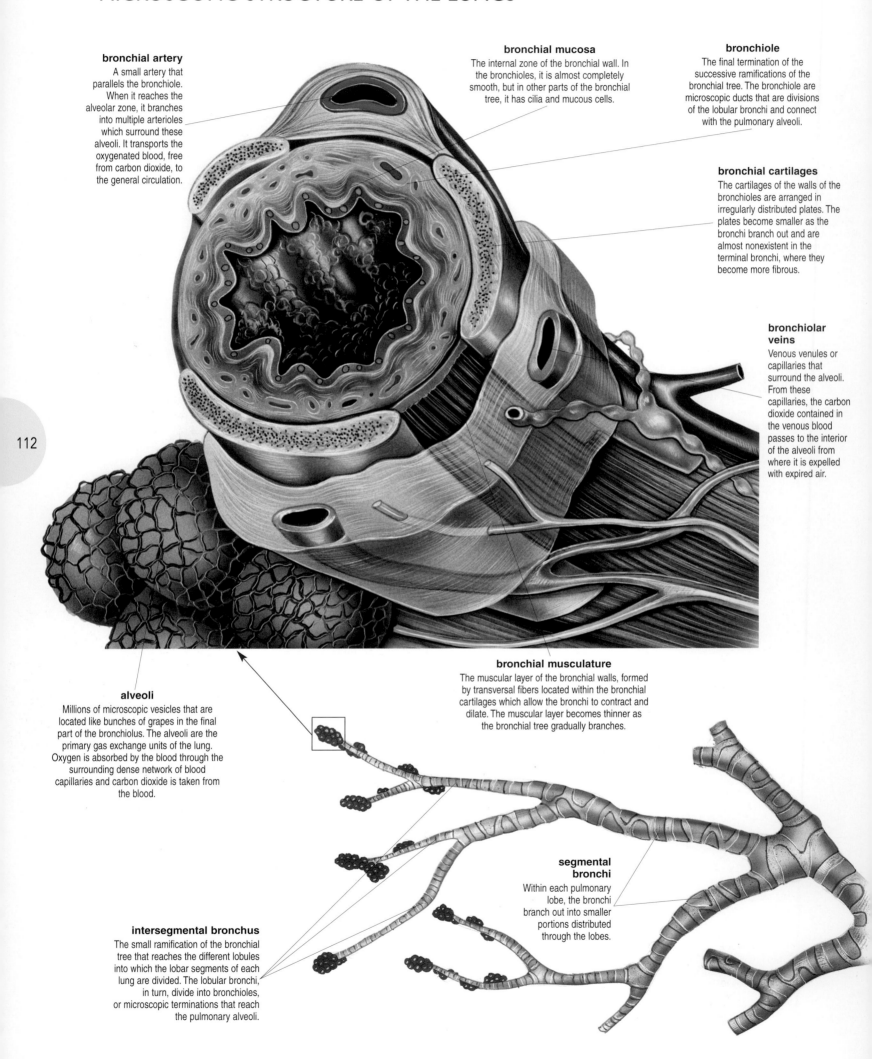

bronchial artery
A small artery that parallels the bronchiole. When it reaches the alveolar zone, it branches into multiple arterioles which surround these alveoli. It transports the oxygenated blood, free from carbon dioxide, to the general circulation.

bronchial mucosa
The internal zone of the bronchial wall. In the bronchioles, it is almost completely smooth, but in other parts of the bronchial tree, it has cilia and mucous cells.

bronchiole
The final termination of the successive ramifications of the bronchial tree. The bronchiole are microscopic ducts that are divisions of the lobular bronchi and connect with the pulmonary alveoli.

bronchial cartilages
The cartilages of the walls of the bronchioles are arranged in irregularly distributed plates. The plates become smaller as the bronchi branch out and are almost nonexistent in the terminal bronchi, where they become more fibrous.

bronchiolar veins
Venous venules or capillaries that surround the alveoli. From these capillaries, the carbon dioxide contained in the venous blood passes to the interior of the alveoli from where it is expelled with expired air.

112

alveoli
Millions of microscopic vesicles that are located like bunches of grapes in the final part of the bronchiolus. The alveoli are the primary gas exchange units of the lung. Oxygen is absorbed by the blood through the surrounding dense network of blood capillaries and carbon dioxide is taken from the blood.

bronchial musculature
The muscular layer of the bronchial walls, formed by transversal fibers located within the bronchial cartilages which allow the bronchi to contract and dilate. The muscular layer becomes thinner as the bronchial tree gradually branches.

segmental bronchi
Within each pulmonary lobe, the bronchi branch out into smaller portions distributed through the lobes.

intersegmental bronchus
The small ramification of the bronchial tree that reaches the different lobules into which the lobar segments of each lung are divided. The lobular bronchi, in turn, divide into bronchioles, or microscopic terminations that reach the pulmonary alveoli.

MEDIASTINUM

MEDIASTINUM
The middle region of the thorax, limited laterally by the lungs, posteriorly by the vertebral column and anteriorly by the sternum. The mediastinum contains the thymus, heart, the thoracic aorta and its branches, the inferior vena cava, the trachea and main bronchi and the esophagus. The portion located in front of the trachea is called the anterior mediastinum and the portion behind the trachea is the posterior mediastinum.

pleura
A double membrane that covers the lungs. The internal layer is attached directly to the pulmonary tissue and is known as the pleural viscera. The external layer, known as the parietal pleura, is attached to the structures surrounding the lungs: the ribs, diaphragm and mediastinum.

esophagus
A cylindrical duct that descends the thoracic cavity, communicating the pharynx with the stomach.

bifurcation of the trachea
The trachea terminates in a bifurcation dividing it into the right and left main bronchi which enter the respective lungs.

left subclavian artery
An artery which originates in the arch of the aorta and ascends to the upper limbs, supplying them with arterial blood.

right pulmonary artery
The pulmonary artery which carries blood from the right ventricle of the heart to the lungs to exchange carbon dioxide for oxygen. It divides into the right and left pulmonary arteries and is the only artery that transports deoxygenated blood.

right inferior lobar bronchus
The main bronchi divide into the lobar bronchi when they enter the lungs. There are three lobar bronchi in the right lung and two in the left lung, which go to the respective lobules.

left pulmonary veins
The two left and two right pulmonary veins carry the oxygenated blood to the left atrium of the heart, from where it passes to the left ventricle and is distributed throughout the body.

horizontal fissure
A sulcus that separates the superior and middle lobes of the right lung.

left pulmonary artery
The pulmonary artery or trunk carries blood from the right ventricle of the heart to the lungs to exchange the carbon dioxide for oxygen. It divides into the right and left pulmonary arteries.

oblique fissure of the right lung
The right lung is divided in three lobes: superior, middle and inferior, separated by fissures. The oblique fissure separates the middle lobe from the inferior lobe.

left atrium
The blood oxygenated in the lungs is transported by the pulmonary veins to the left atrium and from there to the left ventricle.

pleural cavity
The space between the visceral pleura and the parietal pleura. Under normal conditions, it is a virtual cavity, because the two walls are united.

pericardium
A layer of fibrous tissue that encloses the walls of the heart. Like the pleura, it consists of the internal or visceral wall and the external or parietal wall.

costophrenic pleural sinuses
The angle formed between the lungs that meet the internal face of the ribs and the superior face of the diaphragm.

oblique fissure of the left lung
A fissure that separates the superior and inferior lobes of the left lung.

diaphragm
A flat muscle that separates the thoracic and abdominal cavities and supports the lungs.

wall of the left ventricle
The left ventricle of the heart receives the oxygenated blood coming from the left atrium and expels it towards the aorta and thus towards the whole body by contractions of the powerful musculature of its walls.

liver
A large organ located under the right lung from which it is separated by the diaphragm. Its left portions extends below the heart.

inferior vena cava
A large blood vessel that collects the venous blood from the inferior part of the body and transports it to the right atrium. In the subdiaphragmatic region, it receives the hepatic veins.

right pulmonary veins
The right pulmonary veins carry the oxygenated blood from the right lung to the left atrium of the heart. Each lung has two pulmonary veins.

aorta
The aorta originates in the heart and crosses the mediastinum vertically. When it passes through the diaphragm, it becomes the abdominal aorta.

stomach
A large saccular organ which receives food from the esophagus and stores it during the digestive process. The walls of the stomach contain glands which secrete gastric juices that help break down the food. The contraction of the stomach walls mixes the food to aid digestion.

113

THE URINARY SYSTEM

▼ FRONTAL GENERAL VIEW (MALE)

inferior vena cava
A thick vein that ascends the abdomen. It collects the venous blood from the abdominal organs and carries it to the heart.

renal hilum
A fissure in the internal part of the kidneys through which the renal blood vessels and the ureters, ducts that transport urine, enter and leave the kidneys.

renal pelvis
A large cavity which collects urine from the renal hilus. It narrows to form the ureter.

ureters
Two irregularly-shaped ducts which descend from the renal pelvises, crossing the posterior part of the abdominal cavity vertically. They carry urine to the bladder for expulsion.

trigone of the bladder
A triangular formation located in the internal surface of the posterior wall of the bladder. The posterior vertices are the ureteric orifices, which are joined by a muscular rim, and the anterior vertex is the neck of the bladder.

prostate
A glandular organ, found only in men, which is attached to the inferior part of the bladder, surrounding the opening of the urethra. Its glands produce secretions which help to form semen.

suprarenal capsules
Structures which house the suprarenal glands located over the kidneys.

kidneys
Two organs located in the abdominal cavity, behind the peritoneum. The kidneys filter and cleanse the blood and also create the urine from the body's waste products.

renal artery
The right and left renal arteries are branches of the aorta which enter the kidneys through the renal hilus.

renal veins
The right and left renal veins carry filtered blood from the kidney to the inferior vena cava.

abdominal aorta
The portion of the aorta that crosses the abdomen and takes the arterial blood from the heart to the various abdominal organs.

bladder
A hollow organ located in the middle inferior area of the pelvic cavity, which receives the ureters. The walls are formed of musculomembranous tissue. It serves to retain the urine until it is expelled.

orifices of the ureter
Orifices that allow the urine to pass from the ureter to the bladder.

neck of the bladder
Located at the base of the bladder, where the urethra begins, the sphincters of the neck of the bladder control urination.

masculine urethra
A duct that carries the urine from the urinary bladder to be expelled during urination. It is longer than the female urethra because it must descend the length of penis. It is divided into three parts: the prostate urethra, the membranous urethra and the penile urethra.

glans
Enlarged final part of the penis, which is covered by a fold of skin called the prepuce.

external urethral orifice or meatus
The point where urine is expelled from the urethra.

female urethra
A duct that communicates the female bladder with the exterior.

external urethral orifice or meatus
The orifice where urine from the female urethra is expelled.

114

▲ FRONTAL DETAIL (FEMALE)

KIDNEYS

▼ EXTERNAL VIEW

KIDNEYS

Two organs located in the abdominal cavity, behind the peritoneum. The kidneys filter and cleanse the blood and also create urine from the body's waste products. The kidneys are shaped like beans and have an interior concavity which contains the renal hilus.

suprarenal capsule
A structure located on the superior pole of each kidney. It contains the suprarenal glands which secrete hormones such as adrenalin, noradrenaline, glucocorticoids, mineralocorticoids and some sexual hormones.

renal hilum
A fissure in the internal part of the kidneys through which the renal blood vessels and the ureters, or ducts that transport the urine, enter and leave the kidneys.

renal artery
A branch of the abdominal aorta that enters the kidney through the renal hilum. Within the kidney, it branches into multiple arterioles which carry the blood to the functional units of the kidney, the nephrons, where the blood is purified.

ureter
An irregularly-shaped duct which descends from the renal pelvis, crossing the posterior part of the abdominal cavity vertically. It carries urine to the urinary bladder for expulsion.

renal vein
A vein that leaves the kidney through the renal hilum and terminates in the inferior vena cava. It originates in the interior of the kidney through the union of multiple venules which come from the renal functional units, the nephrons, and which carry the blood purified of waste products.

▼ INTERNAL VIEW

renal papilla
The renal papillae are the internal vertices of Malpighi's pyramids, through which the urine is transported to the minor chalices.

renal sinus
A cavity located in the center of the kidney, surrounded by the renal marrow. It contains the renal calyces and the renal pelvis.

renal calyx
The renal calyces are chambers, composed of many smaller or minor calices which receive the urine from the thousands of tiny renal papillae. Each kidney has three major calyces, the superior, medial and inferior.

renal capsule
A fibrous membrane that covers all the external surface of the kidney except for an orifice in the internal face called the renal hilum.

renal marrow
The layer lying just below the renal cortex, which contains the conical structures known as Malpighi's pyramids.

renal cortex
The dense tissue located just below the renal capsule. It contains small points known as Malpighi's corpuscles.

renal pelvis
The renal pelvis is a basin at the base of the kidney which collects urine from the renal calyces and carries it to the ureter.

Malpighi's pyramids
Conical formations consisting of multiple ducts which filter and purify the blood and also manufacture and transport urine along with waste products.

Malpighi's corpuscles
Small structures located in the renal cortex that constitute the fundamental part of the functional units of the kidney, the nephrons, where the blood is filtered.

115

MICROSCOPIC STRUCTURE OF THE KIDNEY

afferent arteriole

A small artery that carries blood to Malpighi's corpuscles for filtration. The arterioles are the continuation of the renal artery branches.

glomerular basal membrane

A porous membrane that lines the tiny blood vessels composing the renal glomerulus. The vessels are essential for the complicated process of glomerular filtration of the blood.

Bowman's capsule

An external membranous layer that covers the renal glomerulus.

NEPHRON

The functional unit of the kidney consists of two well-defined parts: Malpighi's corpuscle and a tubular system made up of the proximal contouring tubules, Henle's loop and the distal contouring tubules. Blood is filtered in the nephron to remove impurities and waste products which are expelled through urine.

RENAL GLOMERULUS

distal contouring tubule

A tubular system which is a continuation of Henle's loop and which returns to the area of Malpighi's corpuscle.

proximal contouring tubule

The first part of the tubular system through which the product of the blood filtration leaves. It is located near Malpighi's corpuscle.

Malpighi's corpuscle

collecting tubule

A tubular duct which receives the distal contouring tubules and collects urine, the end product of renal filtration. This duct crosses the renal marrow through the papillae and terminates in the renal calyces.

efferent arteriole

A small artery that leaves Malpighi's corpuscle carrying the filtered blood. The efferent arterioles unite to form the venules that terminate in the renal vein.

juxtaglomerular apparatus

A complex structure located between the afferent and afferent arterioles, which secretes renin, a substance governing the operation of the kidney and acting as a regulator of blood pressure.

renal glomerulus

A capillary bed formed of a compact bunch of interconnected capillaries, which constitutes the central part of the renal corpuscle or Malpighi's corpuscle. Blood arrives at the glomerulus to be filtered by the basal membrane, retaining proteins, blood globules and other substances which can be used by the body (glomerular filtration).

Malpighi's corpuscle

A structure located in the renal cortex. It receives afferent arterioles carrying blood to be filtered and gives off the efferent arterioles containing filtered blood. It contains a great amount of tiny blood vessels crowded together (renal glomerulus), surrounded by a membrane (Bowman's capsule).

renal papillae

Orifices through which the urine arrives from the collecting tubules to the renal calyces.

Henle's loop

A straight, tubular structure which is a continuation of the proximal contouring tubule. It consists of a descending part which enters the renal marrow and an ascending part that returns to the renal cortex. Its function is to select the products obtained through glomerular filtration.

BLADDER AND URETHRA

▼ INTERNAL VIEW

ureters
Ducts which carry the
urine from the kidneys
to the urinary bladder.

ovary
Two female sexual
glands which are
located on either
side of the body
near the ureters.

deferent duct
The excretory duct of
the testis which joins
the excretory duct of the
seminal vesicle to form
the ejaculatory duct.

peritoneum
A serous layer
located above the
bladder that covers
the abdominal cavity.

bladder
A hollow organ which receives
urine from the ureters. The
bladder walls are formed of
musculomembranous tissue.
The function of the bladders
is to retain the urine until
expulsion.

trigone of the bladder
A triangular area located in
the internal face of the
posterior wall of the bladder.
The posterior vertices are
the ureteric orifices, which
are joined by a muscular
rim, and the anterior vertex
is the neck of the bladder.

ureteral meatus
Orifices through which urine
enters the bladder from the
ureters. The muscular walls of
the bladder form a valve that
prevents a reflux of the urine.

neck of the bladder
The area on the floor of the
bladder where the urethra
begins. The muscular
layer of the bladder wall
thickens to form the
internal urethral sphincter.

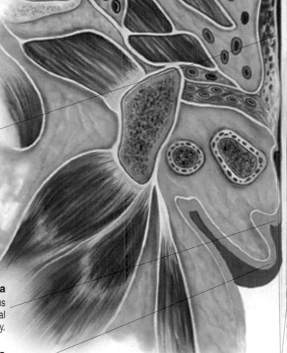

prostate
A gland found only in males,
which manufactures some
components of the seminal fluid.

feminine urethra
The female urethra is
much shorter than the
male one, running
from the neck of the
bladder to the urethral
meatus in the vulva.

prostate urethra
First portion of the masculine
urethra located at the height
of the prostate gland.

membranous urethra
The second part of the masculine
urethra corresponding to the short
portion which crosses the perineum.

penile urethra
The final part of the male
urethra which runs down the
penis and expels urine.

labia menora
One of the two mucocutaneous
folds that border the vaginal
orifice laterally.

scrotum
A cutaneous
sac which
contains the
testicles located
in the anterior
part of the male
perineum.

glans
The final,
enlarged part
of the penis,
covered by a
retractable layer
of skin known
as the prepuce.

penis
A cylindrical organ located
anterior to the scrotum. It
contains the penile urethra
which expels urine.

labia majora
One of the two cutaneous
folds that surround
the smaller lips.

urethral meatus
The external part of
the urethra, through
which urine is expelled.

FEMALE

MALE

117

MALE REPRODUCTIVE SYSTEM

▼ GENERAL VIEW. LATERAL SECTION

peritoneum
A membrane that lines the abdominal cavity and covers the superior part of the bladder and a substantial part of the anterior face of the rectum.

rectouterine pouch or cul-de-sac of Douglas
A fold of the peritoneum that forms a cul-de-sac between the bladder and the rectum.

rectum
The end of the large intestine. It is located behind the prostate, which may become enlarged enough to be felt by rectal examination.

anus
The external orifice of the large intestine.

prostate
A gland located below the bladder. It produces a series of secretions which are mixed with sperm in the urethra to form the seminal fluid that is expelled by ejaculation.

perineum
A region which extends from the posterior part of the scrotal sac to the anus, forming the floor of the pelvic cavity.

erectile tissue
The spongy tissue that surrounds the urethra. It fills with blood during sexual arousal, causing the penis to enlarge and become erect.

sigmoid colon
The final part of the descending colon that ends at the rectum.

urinary bladder
A saccular organ forming part of the urinary system in which the urine is stored before being expelled. It is located above the prostate.

ureter
A duct that communicates the kidney with the bladder.

iliac veins and arteries
The iliac veins and arteries supply the branches which irrigate and collect blood from the reproductive apparatus.

symphysis of the pubis
A joint formed by the union of the pubic bones, constituting the anterior limit of the pelvic cavity.

cavernous bodies of the penis
Cylinders with a spongy structure located in the dorsal part of the penis. They fill with blood during sexual arousal, causing the penis to become enlarged and erect.

urethra
A duct which carries urine from the urinary bladder down the penis to be expelled and semen to be ejaculated from the deferent ducts. It consists of 3 parts: the prostate, membranous and penile urethra.

penis
The external male genital organ, which enlarges and becomes erect during arousal. It houses the urethra which expels urine and deposits semen in the vagina during copulation.

glans
Conical swelling located in the distal end of the penis, which separates it from the balanopreputial sulcus.

urinary meatus
External orifice of the urethra, located in the vertex of the glans. Semen and urine are expelled through it.

scrotum
A saccular structure located in the anterior zone of the perineum, behind the penis, which hangs between both thighs. It contains the testicles.

testicles
Ovoid organs contained in the scrotum. They produce spermatozoa, the male reproductive cells, and the hormones responsible for the appearance of masculine sexual characteristics.

prepuce or foreskin
The skin that covers the distal end of the glans penis. It may retract to leave the glans uncovered.

FEMALE REPRODUCTIVE SYSTEM

▼ GENERAL VIEW. LATERAL CROSS-SECTION

ovary
The two ovaries located on each side of the body are the female sexual glands. They have a double function, producing the ova (female sexual cells) and also manufacturing female sexual hormones. This activity begins in puberty and stops at the menopause.

iliac vein and artery
The iliac vein and artery originate in the inferior vena cava and the abdominal aorta respectively and descend to the pelvic area where they ramify into the branches which irrigate and collect blood from the reproductive apparatus.

ureter
A duct that communicates the kidney with the urinary bladder. In this area, it passes close to the Fallopian tubes and the uterus.

sigmoid colon
The distal part of the descending colon that ends at the rectum.

Fallopian tubes
The two Fallopian tubes communicate the ovaries with the superior part of the uterine cavity. They transport the egg to the uterus after it is released by the ovary.

broad ligament
A large ligament that unites the female uterus and other genital organs with the walls of the pelvic cavity.

peritoneum
A membrane that lines the abdominal cavity and covers the superior part of the bladder and a substantial part of the anterior face of the rectum.

rectouterine pouch or **cul-de-sac of Douglas**
A fold of the peritoneum that forms a cul-de-sac behind the uterine cavity.

bladder
A saccular organ in which urine is stored before being expelled.

rectum
The final portion of the digestive system located immediately behind the uterus, whose posterior wall it supports.

uterus
A hollow organ formed of thick muscular walls, located in the medial part of the pelvic cavity and divided into three sections: the wider uterine body, the medial uterine isthmus and the thinner uterine neck. The uterus lodges the fertilized egg as it grows.

symphysis of the pubis
A joint formed by the union of the pubic bones, constituting the anterior limit of the pelvic cavity.

mount of Venus or **pubis**
A projection in the skin located just under the skin in the superior part of the vulva.

urethral orifice
A small orifice located below the clitoris and above the vaginal orifice through which urine is expelled from the urethra.

clitoris
An erectile organ located in the vertex of union of the labia majora, partially covered by a cutaneous fold.

labia majora
One of the two cutaneous folds surrounding the labia minora.

labia minora
One of the two mucocutaneous folds that border the vaginal orifice laterally.

vaginal orifice
The external orifice of the vagina, which is covered by a fine membrane called the hymen which is broken when sexual relations are initiated.

vulva
The external female genital organs.

anus
The external orifice of the large intestine.

perineum
A region which extends from the inferior extreme of the vulva to the anus, forming the floor of the pelvic cavity.

vagina
A duct formed of muscle and membrane that originates in the cervical neck and communicates with the exterior through the vagina orifice. Its function is to receive the masculine penis during copulation. It can expand enormously to accommodate the fetus during childbirth.

119

THE PENIS

▼ ANTERIOR VIEW

base of the penis
Also called the root of the penis. The area where the penile shaft connects with the inferior part of the abdomen.

penile shaft
The central, cylindrical part of the penis.

glans
A conical swelling located in the distal end of the penis. Its widest part is the corona of the glans. The urethral meatus opens in its distal end.

scrotum or **scrotal sac**
A saccular structure which hangs between both thighs located in the anterior zone of the perineum, behind the penis.

pubic hair
Thick, curly hair which covers the inferior part of the abdomen above the penis.

PENIS
The external male genital organ, which enlarges and becomes erect during arousal. It houses the urethra which expels urine and allows semen to be deposited in the vagina during copulation. It consists of the base, the shaft and the distal extreme or glans.

prepuce or **foreskin**
The skin that covers the glans. It may retract to leave the glans uncovered.

balanopreputial sulcus
The sulcus that delimits the crown of the glans and separates it from the penile shaft.

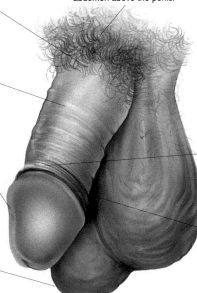

▼ POSTERIOR VIEW

coronal sulcus
A sulcus that crosses the surface of the glans from the urethral meatus to the frenulum. It is uniformly smooth and uniform except in the anterior face.

urethral meatus
The external orifice of the urethra located in the vertex of the glans. It expels urine and ejaculates semen.

glans
Conical swelling located in the distal end of the penis from which it is separated by the balanopreputial sulcus.

frenulum
A cutaneous fold that joins the prepuce and the glans in the posterior face of the penis.

penile shaft
The central, cylindrical part of the penis.

prepuce
The skin that covers the distal end of the glans penis. It may retract, leaving the glans uncovered.

VULVA

▼ FRONTAL VIEW

mount of Venus or **pubis**
An area located in the superior part of the vulva. At puberty, it is covered by thick, heavy pubic hair.

clitoris
An erectile organ located in the vertex of the two labia minora. It is formed of erectile tissue that fills with blood during sexual stimulation.

frenulum of the clitoris
Cutaneous folds of the labia minora in its anterior part which unite in the anterior part of the clitoris in a way that imitates the male frenulum.

urethral orifice
A small orifice, located below the clitoris and above the vaginal orifice, through which urine is expelled from the urethra.

opening of Bartholin's glands
A pair of glands located between the vagina and the vulva which become lubricated when stimulated, thus facilitating sexual intercourse. They are also called the greater vestibular glands.

vaginal orifice
The external orifice of the vagina, which is covered by a fine membrane, called the hymen, which is broken when sexual relations are initiated.

posterior commissure or **fourchette of the vulva**
The angle formed by the union of the posterior parts of the labia majora.

VULVA
The external female genital organs. Located in the inferior zone of the abdomen between the thighs, it constitutes the visible, exterior part of the female reproductive apparatus.

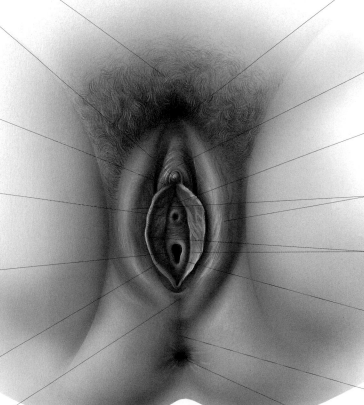

anterior commissure of the labia majora
The angle formed by union of the posterior parts of the labia majora.

prepuce of the clitoris
A cutaneous fold formed by the anterior part of the labia minora, which covers the clitoris in a similar way to the male prepuce.

vestibule
The vestibule of the vagina is the name given to the cleft or space which is surrounded by the labia minora.

labia majora
One of the two cutaneous folds that surround the labia minora. Their anterior areas are a rosy color and the posterior areas are somewhat darker.

labia minora
One of the two mucocutaneous folds that border the vaginal orifice laterally.

hymen
An incomplete membrane that partially blocks the vaginal orifice, usually broken when sexual relations are initiated. It can adopt very different forms.

perineum
An area located between the thighs that extends from the posterior commissure to the anus.

anus
The external orifice of the rectum and the posterior limit of the perineum.

OVARIES, TUBES AND UTERUS

▼ EXTERNAL AND INTERNAL ANTERIOR VIEW

uterus
In the pre-pregnant state, the uterus is the shape and size of a small pear, and is about one inch thick. It contains the fertilized ovum.

Fallopian tube or uterine tube
The Fallopian tubes are the ducts that communicate the ovaries with the uterus. Their function is to collect the ovum, once released from the ovary, and transport it to the uterus. They consist of three portions: the isthmus, the ampulla and the infundibulum. Their walls are formed by an external serous tunica, two medial layers of smooth musculature and one internal mucous layer.

mesosalpinx of the broad ligament
membrane that connects the pavilion of Fallopian tube with uterus, ensuring they do not become separated and no released ovum is lost.

cervical neck
The lowest, narrowest part of the uterus, which ends at the vagina.

body of the uterus
The superior, wider part of the uterus which houses the uterine cavity.

endometrium
A membranous lamina that lines the uterine cavity. It houses the fertilized egg and is shed during the menstrual cycle when the woman is not pregnant.

myometrium
Three thick layers of smooth muscle that form the greatest part of the uterine walls.

isthmus
A transitional area that unites the body of the uterus with the neck.

uterine ostium
The orifice located in the superior angle of the uterus which connects it with the uterine or Fallopian tube.

isthmus
The narrower, middle part of the Fallopian tube that joins the uterine wall.

ampulla
The mid-region of the Fallopian tube. It has thin walls and virtually no muscle.

infundibulum and folds of uterine tube
The external part of the uterine tube which ends in the fimbriae. It is joined to the ovary by the membranes of the broad ligament. Its function is to capture the ova released by the ovary.

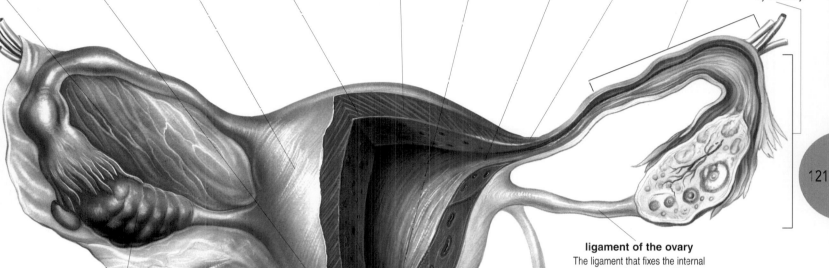

121

ligament of the ovary
The ligament that fixes the internal portion of the ovaries with the superior angles of the uterus.

external cervical orifice
The portion of the cervical neck that communicates directly with the vaginal cavity.

vaginal fornix
The anterior and posterior recesses into which the upper vagina is divided and which are formed by the protrusion of the cervix into the vagina.

transversal mucous folds
A series of circular sulci that cross the internal face of the vaginal mucosa. They correspond to the marks left by the circular muscular fibers.

tunica mucosa layer of the vagina
A fine membrane that lines the surface of the vagina. It is a continuation of the mucosa of the uterus.

tunica muscularis of the vagina
The vaginal walls are equipped with two layers of smooth muscular fibers, one longitudinal and the other circular.

tunica serosa of the vagina
A thin layer of connective tissue that covers the vaginal walls.

vagina
A duct composed of muscle and membrane that originates in the cervical neck and communicates with the exterior through the vaginal orifice. It receives the masculine penis during copulation. It can expand enormously to accommodate the fetus during childbirth.

ovary
The ovaries are two pink, ovoid glands located on both sides of the female pelvic cavity. They are the feminine sexual glands and their activity begins at puberty and stops with the onset of menopause. They produce the ova, the female sexual cells and hormones.

round ligament
A cord that connects the wall of the uterine body with the anterior wall of the abdomen, passing through the inguinal canal.

mesometrium of broad ligament
Prolongation of the peritoneal layer that separates the pelvic and abdominal cavities. It is arranged in front, above and behind the uterus and covers the tubes and the ovaries.

transverse muscle of the vagina
A branch of the deep transverse muscle that goes from the cervical neck to the ischiopubian branches.

THE BREASTS

breasts
Two hemispheric structures that contain the female mammary glands and are located in the anterosuperior zone of the female chest. They serve to secrete the breast milk that nourishes the newborn baby.

mammary areola
An area of darker, furrowed skin located around the nipple.

nipple
A protuberance or papilla located in the center of the mammary areola, which serves as the external orifice for the ducts that carry the breast milk.

Morgagni's tubercles
Small protuberances which cover the mammary areola, giving it a rough, granular aspect. Each contains a sebaceous gland which may contain a hair.

submammary fold
A semicircular crease that lies under each breast where it joins the thorax.

122

▲ LATERAL VIEW OF THE RIGHT BREAST   ▲ FRONTAL VIEW OF THE RIGHT BREAST

MAMMARY GLAND

A lobular, glandular system which produces milky secretions after childbirth and transports them to the nipple by a series of canaliculi.

---

subcutaneous adipose layer
Abundant adipose or fatty tissue located below the skin of the mammary area, which it surrounds and protects.

greater pectoral muscle
Muscle that covers the superoanterior area of the thorax underneath the breast.

gland lobule
The mammary gland is divided into multiple lobules which constitute the functional units.

suspensory ligaments or ligaments of Cooper
Located in the posterior face of the mammary gland, the suspensory ligaments unite the breast to the aponeurosis of the greater pectoral muscle located behind it.

lactiferous sinus
Small dilations located in the final part of each of the lactiferous ducts.

lactiferous ducts
Winding ducts that communicate each of the gland lobules with the exterior through the nipple, carrying the breast milk.

ribs
Saggital section of the ribs which support the superior pectoral muscle and the breast.

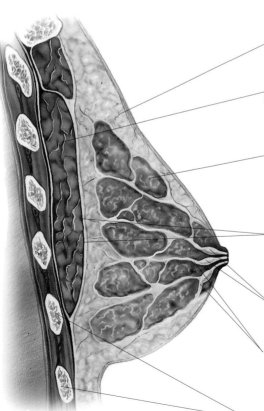

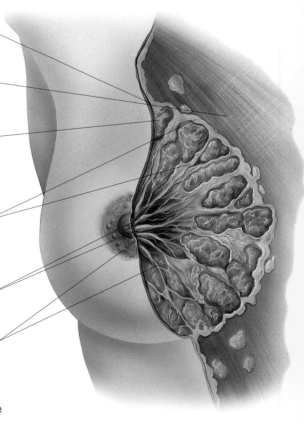

▲ SECTIONS OF THE MAMMARY GLAND

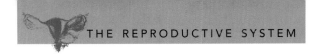

ABDOMEN OF A PREGNANT WOMAN

▼ LATERAL CROSS-SECTION

stomach
The compression suffered by the stomach and the intestinal loops during the last months of pregnancy causes frequent regurgitations which may cause some discomfort such as heartburn or nausea.

liver
During pregnancy, the liver is compressed by the enlarged uterus, which may result in difficulties in draining the gallbladder to the intestine.

fetus
The fetus grows from an embryo to the moment of birth. During the later stages of pregnancy, the fetus is perfectly formed.

vertebral column
The increase in abdominal volume during pregnancy moves the center of gravity forwards, creating a greater anterior curvature of the vertebral column, or lordosis, which causes backaches.

intestinal loops of the small intestine
These are also compressed by the increased size of the uterus.

uterus
A hollow cavity with thick, muscular walls that progressively expands during pregnancy to accommodate the growing fetus, which gradually occupies nearly all the abdominal cavity of the pregnant woman.

rectum
The final part of the digestive system that communicates with the exterior through the anus. The lack of intestinal mobility due to the pressure of the uterus leads to constipation in some pregnant women.

breasts
During pregnancy, the breasts increase progressively in size and the pigmentation of the areola and the nipple darkens.

vagina
A duct formed of muscle and membrane that originates in the cervical neck and communicates with the exterior through the vaginal orifice. It can expand enormously to accommodate the fetus during childbirth.

cervical neck
The cervical neck remains firmly closed during pregnancy and only becomes dilated at the moments preceding childbirth, in order to allow the passage of the fetus towards the exterior.

bladder
A cavity in which the urine is stored before being expelled. In the final phase of pregnancy, the pressure exerted by the head of the fetus on the bladder causes frequent urination accompanied with discomfort.

placenta
A structure that forms at the end of the first month of pregnancy. It is rich in blood vessels which are attached to the uterine wall and to the fetus by the umbilical cord. It supplies blood and nutrition to the fetus throughout gestation.

umbilical cord
A tubular cord of variable length which unites the center of the placenta with the fetus. It contains two arteries and a vein which carry arterial blood to and venous blood from the fetus.

umbilical herniation
The intraabdominal pressure caused by the increased size of the uterus, causes, in many women, a swelling of the umbilicus, which is, in fact, an abdominal hernia through the umbilicus.

amniotic sac
From the beginning of the pregnancy, a membrane forms around the embryo in the shape of a sac which surrounds the fetus during gestation. The sac contains amniotic fluid and only breaks at the moment of birth.

amniotic fluid
A fluid contained inside the amniotic sac, where the fetus is located. It is composed mainly of water, but also contains epithelial cells, feta urine, salts and enzymes. It serves is to protect the fetus.

symphysis of the pubis
The articulation of the two pubic bones in the anterior part of the pelvic waist, which, during the final phase of pregnancy and in childbirth, allows a small widening of the channel through which the fetus passes.

123

COMPONENTS OF THE BLOOD

Diagram of a blood capillary with the components of the blood

blood plasma
The fluid component fluid of the blood, representing 55-60% of its total volume. Plasma has a yellowish color and is composed mainly of water, but contains numerous substances, such as proteins, minerals, sugars, enzymes and vitamins.

cellular elements of the blood
The blood cells represent 40-45% of the blood volume and are of three types: red blood cells (or erythrocytes), white blood cells (or leukocytes), and platelets (or thrombocytes).

124

monocyte
A large, bluish blood cell, whose primary function is to defend the body against prolonged or chronic infections.

lymphocyte
A bluish cell with a single, large nucleus. There are two types of lymphocytes: T cells, which defend the body against viruses and provoke some allergic reactions, and B cells which create antibodies and synthesize some proteins of the immune system.

platelets
Tiny blood cells which, like the red blood cells, have no nucleus. They promote the coagulation of the blood to prevent blood loss through hemorrhages and thus maintain hemostasis.

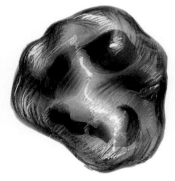

neutrophil
A leukocyte which contains several nuclei. Its cytoplasm contains grains that give the cell a violet color. Neutrophils destroy bacteria by a process known as phagocytosis.

basophil
A leukocyte with several nuclei whose cytoplasm contains grains that give it a purplish color. It is part of the body's defense system.

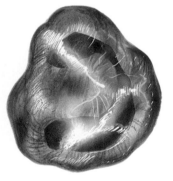

eosinophil
One of the leukocytes that, together with neutrophils and basophils, are called granulocytes. Leukocytes have several nuclei and their cytoplasm is a yellowish-red color. They serve is to defend the body by blocking the antigen-antibody complexes that form when foreign bodies penetrate the human organism.

red blood cells
Rounded blood cells with no nucleus which can change shape to adapt to narrow blood capillaries. Blood can contain more than 5 million per mm^2. They contain hemoglobin, which transports oxygen to the cells.

white blood cells
Blood cells, which unlike the red cells, etc. have a nucleus. Their main function is to defend the body against the infections caused by foreign germs. Also known as leukocytes, they can cross the pores of the blood vessels to reach any focus of infection. There are several types of white blood cells: granulocytes, lymphocytes and monocytes; granulocytes may be mononuclear or polynuclear (neutrophils, basophils and eosinophils).

SPLEEN

splenic artery

RA branch of the celiac trunk which originates in the abdominal aorta and emits arterial branches to the liver, stomach and spleen. The splenic artery emits small branches that irrigate the pancreas.

splenic vein

A vein formed by the union of several venous branches which leave the spleen. It joins the superior and inferior mesenteric veins to form the portal vein that goes to the liver.

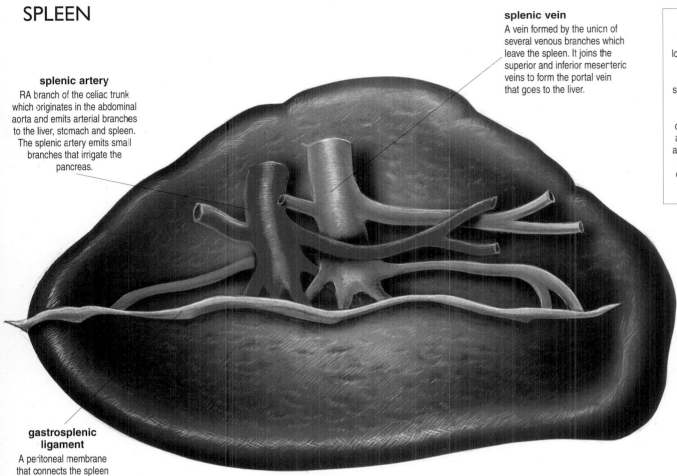

SPLEEN

The spleen is a lymphatic organ located in the superior left quadrant of the abdominal cavity or left hypochondrium, behind the stomach and below the diaphragm. Its function is to destroy the red blood cells after they have completed their mission. It acts as a reservoir of blood cells and may also produce them. It is also part of the body's immunologic system, contributing to the manufacture of antibodies.

gastrosplenic ligament

A peritoneal membrane that connects the spleen and the greater curvature of the stomach.

125

▲ EXTERNAL VIEW

splenic pulp

There are two types of splenic pulp. The red pulp, which forms 75% of the capacity of the spleen, contains massive amounts of red blood cells which give it a deep red color. The white pulp is formed of lymph tissue. The red pulp contains an intricate arterial network which forms the splenic sinuses.

trabecular arteries and veins

Numerous vascular branches which constitute an intrasplenic network. They ensure that the spleen is supplied with the abundant circulation it needs.

splenic capsule

A cortex, formed of membranous connective tissue, which covers the spleen. It emits prolongations that divide the organ into lobes or lobules.

hilium of the spleen

A fissure in the internal face of the spleen, which allows the splenic artery and vein to reach the spleen.

▲ INTERNAL VIEW

GENERAL VIEW OF THE GLANDULAR SYSTEM

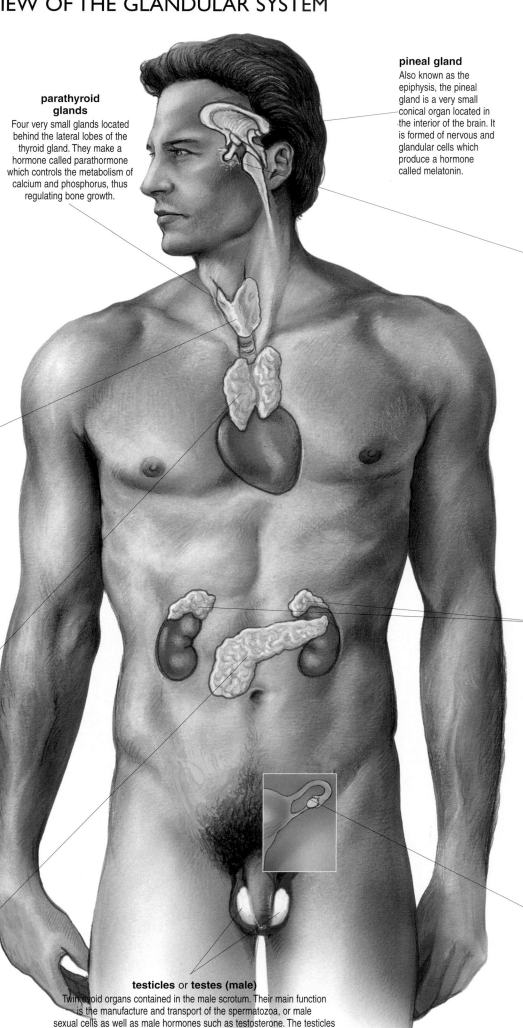

ENDOCRINE SYSTEM

A complex set of interrelated glands that regulate the different metabolic functions of the body and the intensity of the chemical activity of the various cells through the action of substances called hormones. Hormones are chemical substances secreted by a cell or group of cells, which act on the functioning of other cells of the body.

parathyroid glands

Four very small glands located behind the lateral lobes of the thyroid gland. They make a hormone called parathormone which controls the metabolism of calcium and phosphorus, thus regulating bone growth.

pineal gland

Also known as the epiphysis, the pineal gland is a very small conical organ located in the interior of the brain. It is formed of nervous and glandular cells which produce a hormone called melatonin.

pituitary gland or hypophysis

A small, single, ovoid gland located in the interior of the skull within the cavity of the sphenoid bone known as the sella turcica. It is formed of two parts. The anterior, or adenohypophysis, makes hormones that regulate other glands, such as the thyroid (thyroid stimulating), adrenal cortex (adenocorticotropic) or sexual glands (follicle stimulating and luteinizing), as well as a growth-regulating hormone. The posterior or neurohypophysis produces hormones that regulate the operation of the kidney (vasopressin) and childbirth and breastfeeding in women (oxytocin).

thyroid

A gland located in the neck, in front of the trachea. It is divided into right and left lobes which are connected by a narrow area called the thyroid isthmus. It produces hormones called thyroxine and triiodothyronine which regulate the basal metabolism and the maturation of the nervous system.

thymus

A gland located in the thoracic cavity, behind the sternum. It consists of right and left lobes and contains lymphoid cells called thymocytes, which produce antibodies to defend the body against foreign substances by stimulating the blood lymphocytes. This function is especially important in young children, but seems to be less so in adults.

adrenal or suprarenal glands

Twin glands located in the superior pole of the kidneys. They are composed of two parts, the peripheral or suprarenal cortex and the central or suprarenal marrow. The suprarenal cortex produces mineral-corticoids that maintain the balance between fluids and different minerals of the body, glucocorticoids which regulate the metabolism of glucose, fats and proteins, and small amounts of sexual hormones. The suprarenal marrow makes hormones called adrenalin and noradrenaline that act on the nervous system and regulate stress.

pancreas

An organ located in the superior part of the abdominal cavity which has both exocrine and endocrine functions. The exocrine function consists of the secretion of pancreatic fluids to the duodenum to aid the digestion of foods. The endocrine function is the secretion of insulin and glucagon, hormones which go to the blood, and regulating the uptake of glucose, the main cell nutrient.

ovaries (female)

Twin glands located inside the female pelvic cavity that remain inactive until puberty. Once activated, they have a double function: to manufacture the female sexual cells or ova, that are released in each ovarian cycle and to manufacture estrogen and progesterone, the female hormones, which determine the appearance of female characteristics and regulate the menstrual cycle. The ovaries communicate with the rest of the sexual organs through the Fallopian tubes.

testicles or testes (male)

Twin ovoid organs contained in the male scrotum. Their main function is the manufacture and transport of the spermatozoa, or male sexual cells as well as male hormones such as testosterone. The testicles communicate with the rest of the genital organs through the spermatic ducts.

HYPOPHYSEAL CONTROL

PITUITARY GLAND

Also known as the hypophysis. A small, single, ovoid gland, located in the interior of the skull within the cavity of the sphenoid bone known as the sella turcica. In a complex system, governed by a superior structure called the hypothalamus, the pituitary gland regulates the hormonal secretions of the other glands in the body.

neurosecretory cells

The neurosecretory cells are located in the hypothalamus. They secrete hormones called neurosecretory substances, which are carried by the hypothalamus-hypophysial portal system to the adenohypophysis, where they regulate the glandular cells which stimulate the production of hormones. They are called hormone regulation factors.

adenohypophysis

The anterior lobe of the pituitary gland. It contains a series of glandular cells that, when stimulated by the neurosecretory substances produced in the hypothalamus, secrete different types of hormones which stimulate the other glands of the body, such as the thyroid, suprarenal or sexual glands that act on specific tissues, such as the growth hormone.

stimulating hormone of the melanocytes

A hormone whose production is governed by the pineal gland. It stimulates the cells of the dermis to produce melanin, a pigment which colors the skin.

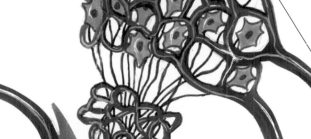

hypothalamus

A nervous organ located in the base of brain, in the floor and lateral walls of the third ventricle. It contains numerous neurological centers which regulate activities such as sight, sleep and, through nervous stimuli and hormonal secretions, the operation of the hypophysis.

hypophysial stem

The hypophysis is united to the hypothalamus of the brain through a pedicle or hypophysial stem, which contains some nervous terminations and a dense network of blood vessels that unite both structures.

antidiuretic hormone or vasopressin

A hormone produced by neurohypophysis that regulates the amount of water the kidney reabsorbs, thereby controlling the amount of urine produced.

corticotropin

A hormone produced by the adenohypophysis which stimulates the cortex of the suprarenal glands, causing them to secrete glucocorticoids and mineral-corticoids.

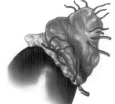

oxytocin

A hormone produced by neurohypophysis that acts on the uterine musculature, causing contractions during childbirth, and on the mammary glands and facilitating the production of breast milk.

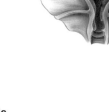

thyrotropin

Adenohypophysial hormone which stimulates the thyroid gland to produce the thyroid hormones, thyroxine and triiodothyronine.

follicle-stimulating hormone

An adenohypophysial hormone that begins to secrete during puberty and acts on the ovaries, stimulating the development of the follicles. In men, it acts on the testicles, initiating the production of spermatozoa.

luteinizing hormone

A hormone produced by adenohypophysis which complements the follicle-stimulating hormone as well as regulating ovulation in women and the production of testosterone in men.

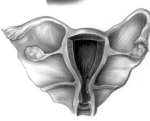

neurohypophysis

The posterior lobe of the hypophysis. It is attached to the hypothalamus by nerve fibers from the nervous centers of the hypothalamus. It produces two types of hormone: the antidiuretic hormone or vasopressin and oxytocin.

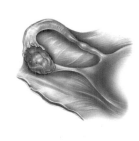

growth hormone

The growth hormone is an adenohypophysial hormone that does not act on another gland, but mainly on growing tissues, where it increases the synthesis of proteins and facilitates the production of energy from fats.

NERVOUS SYSTEM

▼ DORSAL GENERAL VIEW

ORGANIZATION OF THE NERVOUS SYSTEM

The nervous system is composed of a set of interconnected organs whose complex operation allows it to control to the rest of the body's systems. There is a central nervous system, formed of the brain (cerebrum, mesencephalon, medulla oblongata and cerebellum) and the spinal marrow, and a peripheral nervous system, formed of the ganglia and the nerves.

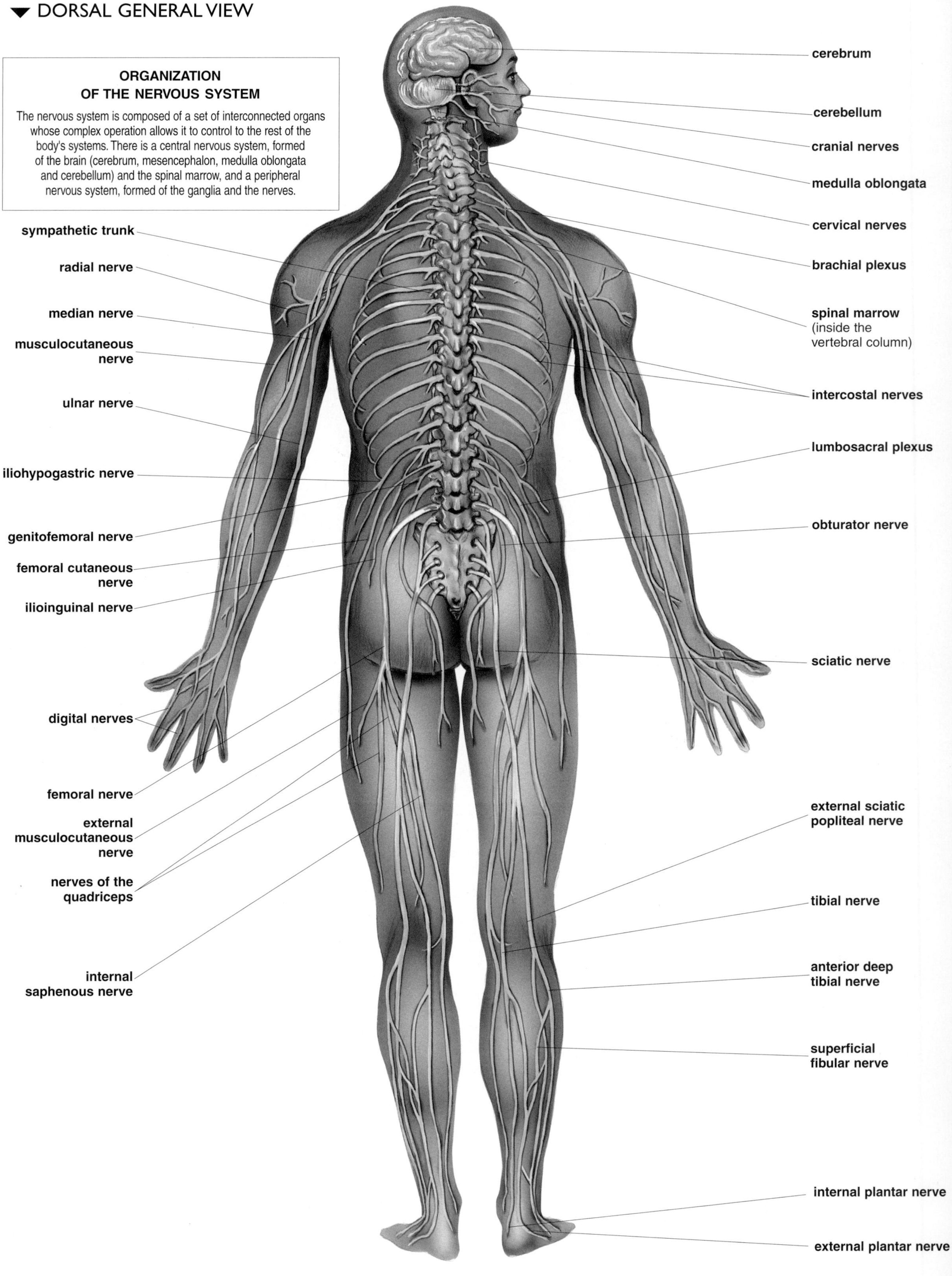

sympathetic trunk

radial nerve

median nerve

musculocutaneous nerve

ulnar nerve

iliohypogastric nerve

128

genitofemoral nerve

femoral cutaneous nerve

ilioinguinal nerve

digital nerves

femoral nerve

external musculocutaneous nerve

nerves of the quadriceps

internal saphenous nerve

cerebrum

cerebellum

cranial nerves

medulla oblongata

cervical nerves

brachial plexus

spinal marrow (inside the vertebral column)

intercostal nerves

lumbosacral plexus

obturator nerve

sciatic nerve

external sciatic popliteal nerve

tibial nerve

anterior deep tibial nerve

superficial fibular nerve

internal plantar nerve

external plantar nerve

AUTONOMIC NERVOUS SYSTEM

AUTONOMIC NERVOUS SYSTEM

The part of the nervous system that regulates the internal activity of the body, controlling the operation of organs such as the heart, blood vessels, the intestines, the kidneys and the different glands. All of the organs act totally independently of the conscious will of the individual. The automatic nervous system consists of two parts: the sympathetic system, which prepares the organism for stressful situations requiring alertness, and the parasympathetic system.

ocular branches of the sympathetic trunk
The sympathetic system controls the ciliary musculature of the eye, causing an expansion of the pupil.

salivary branches of the sympathetic trunk
The sympathetic system acts on the salivary glands by diminishing the salivary secretion. Thus, in situations of fear, in which the sympathetic system is activated, the mouth becomes dry.

sympathetic trunk
A nervous chain formed by a succession of ganglia, located on either side of the vertebral column, which goes from the cervical zone to the lumbar one. The ganglia receive nerve fibers from the spinal marrow, which are connected with the superior centers of control located in the hypothalamus. The ganglia give off nerve signals which reach the different viscera.

cardiac branches of the sympathetic trunk
The sympathetic system acts on the heart, increasing the frequency and force of the heartbeat and expanding the coronary arteries.

pulmonary branches of the sympathetic trunk
The sympathetic system causes an expansion of the trachea and the bronchi, allowing more air to reach the lungs.

aortic branches of the sympathetic trunk
The sympathetic system controls the operation of the blood vessels, acting to contract them and thus increase blood pressure.

splanchnic branches of the sympathetic trunk
The sympathetic system reaches the stomach and other intestinal organs through the splanchnic branches, causing a reduction of the peristaltic movements, slowing intestinal transit and increasing the muscular tone of the sphincters. The branches act on the kidney, causing a reduction of urine production.

vesical and prostate branches of the sympathetic trunk
The sympathetic system acts on the bladder, contracting its sphincter.

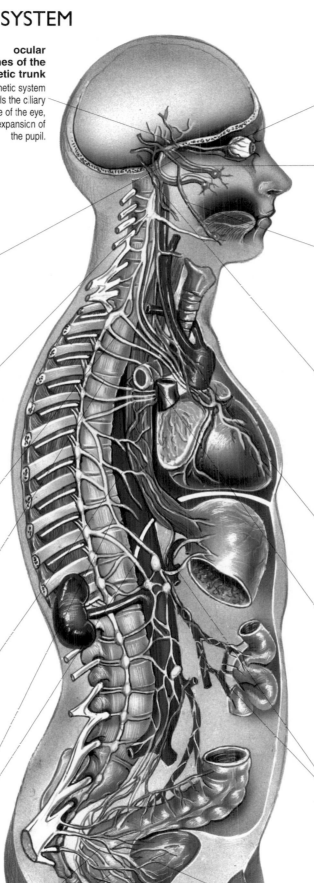

oculomotor nerve
The third of the cranial nerves, which contains parasympathetic fibers which control the ciliary musculature of the pupil. The parasympathetic system produces a contraction in the pupils.

facial nerve
The seventh cranial pair, containing some parasympathetic fibers which stimulate the lacrimal, salivary and nasal secretions.

glossopharyngeal nerve
The ninth cranial pair, which carries some parasympathetic fibers that go to the parotid glands, controlling their secretion.

vagus nerve
Tenth of the cranial pairs, which originates in the medulla oblongata and descends through the neck, thorax and abdomen, sending nervous branches to the different organs of the zones. Most of the fibers of the parasympathetic system pass through the vagus, although some also pass through other cranial pairs.

cardiac branches of the vagus nerve
The parasympathetic system acts on the heart, diminishing the frequency and forces of the heartbeat and contracting the coronary arteries.

pulmonary branches of the vagus nerve
The parasympathetic system contracts the tracheal and bronchial musculature of the lung.

intestinal branches of the vagus nerve
The parasympathetic system acts on the stomach and intestine through the intestinal branches, increasing peristaltic contractions and accelerating intestinal transit, while simultaneously relaxing the sphincters.

vesical and prostate branches of the vagus nerve
The parasympathetic system relaxes the sphincter of the bladder and stimulates the sexual organs.

129

SYMPATHETIC SYSTEM
(represented in yellow)

PARASYMPATHETIC SYSTEM
(represented in green)

NEURONS

NEURON

The fundamental cell of the nervous tissue, which receives and manufactures information and generates and transmits responses. It is composed of a cellular body as well as prolongations which connect with other neurons and conduct the nervous impulses. According to the shape of the cellular body, neurons can be spherical, polyhedral, star-shaped or conical. Depending on the prolongations, they may be unipolar, bipolar or multipolar. The neurons do not regenerate or reproduce, meaning the body's total amount of neurons is fixed from a very young age.

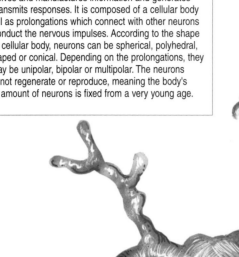

blood capillary
The nervous cells, due to their specific function, have high levels of metabolism. For this reason, their tissue is very rich in blood capillaries, to which the glial cells are attached.

dendrites
Irregularly shaped prolongations of the cytoplasm of the cellular body. They conduct the nervous impulses generated by other neurons to the cellular body. A neuron can have many dendritic prolongations.

cellular body
The central part of the neuron which contains the nucleus. It is surrounded by cytoplasm and the remaining intracellular corpuscles (including the Golgi apparatus and the mitochondria).

axon
A prolongation of the cellular body with a differentiated structure. The axons constitute most of the nerve fibers and the nerves of the organism. They serve is to conduct the nervous impulse generated in the cellular body to other neurons. Generally, each neuron has a single axon and its length is usually much greater than that of the dendrites.

glial cells
Cells that form the support tissue, the structural support of the central nervous system, by mixing with the neurons. There are various types, including astrocytes, oligodendrocytes and ependymal cells. These cells emit vascular prolongations that adhere to the blood capillaries and the neurons.

Ranvier's nodes
Areas of the axon are not covered by the myelin sheath.

myelin sheath
Axons are covered by a sheath made of a substance called myelin composed of lipoproteins and produced by some glial cells. It supports the axon and increases the speed at which the nervous impulse is transmitted.

Schwann cells
Schwann cells are similar to oligodendrocytes in function. They provide myelination to axons in the peripheral nervous system and also have phagocytotic functions.

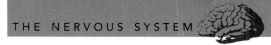

STRUCTURE OF A NERVE. SYNAPSE

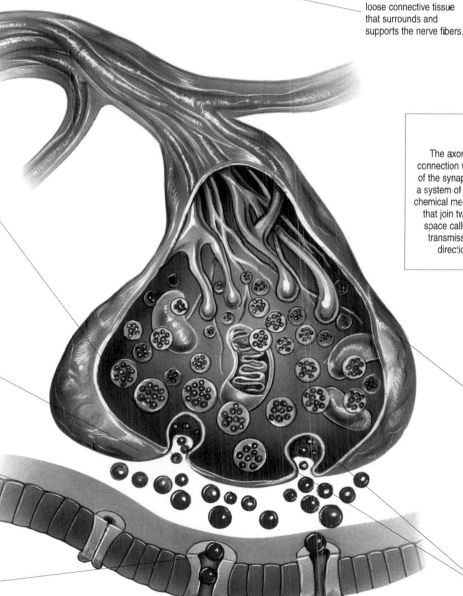

nerve fibers or axons
Prolongations of the neuronal bodies which transport the nervous impulses generated in the central nervous system through the nerves to all parts of the body. Some axons are covered by a myelin sheath and others are not.

nervous fascicle
Groupings of nerve fibers or axons that run through the nerve.

epineurium
A layer of dense connective tissue which covers the nerve.

ganglion
A structure formed by groupings of neuronal bodies located in the near the spinal marrow covered by a layer of connective tissue. The ganglia are intermediate stations in the transmission of the nervous impulse from the central nervous system to the periphery.

perineurium
A layer of dense connective tissue located in the interior of the nerve, covering the nervous fascicles.

blood capillaries
The nervous cells, due to their specific function, have high levels of metabolism. For this reason, their tissue is very rich in blood capillaries, to which the glial cells are attached.

endoneurium
A structure formed of loose connective tissue that surrounds and supports the nerve fibers.

nerve
A structure which carries the axons or nerve fibers of the neuronal bodies of the central nervous system, or ganglia, through the body. Some nerves transmit motor orders (motor nerves), others collect sensory information (sensory nerves) and others carry out both functions (mixed nerves).

131

neurotransmitters
Neurotransmitters are substances which, when released in the synapse, are captured by the receivers located in another cell, producing electrical changes which generate a nervous impulse. There are different types of neurotransmitters specialized in transmitting one or another type of nervous impulses, such as acetycholine, dopamine and noradrenaline. Once these neurotransmitters have acted, they are destroyed or reincorporated in the terminal button.

SYNAPSE

The axons of a neuron establish a connection with other neurons by means of the synapse, which, rather than being a system of physical contact, act through chemical mediators. Between the surfaces that join two neurons, there is a small space called the synaptic space. The transmission is always in the same direction and is not reversible.

presynaptic membrane
A membrane located on the surface of the terminal button close to the neuron with which it connects. It contains pores through which the neurotransmitters reach the synaptic space.

terminal button
A widening of the axon, located in its terminal portion, exactly in the place where it makes contact with another neuron. It contains the elements that allow this contact.

postsynaptic membrane
A membrane located on the surface of the neuron which receives the synaptic connection. It contains specific neuroreceptors.

neuroreceptors
Receivers located in the postsynaptic membrane which capture the signals induced by the neurotransmitters and turn them into electrical signals which generate nervous impulses.

synaptic vesicles
Sacs or vesicles which contain the substances or neurotransmitters which make the chemical connection between two neurons. They open following the orders of the electrical impulses transmitted by the axons.

CEREBRUM

▼ INFERIOR VIEW

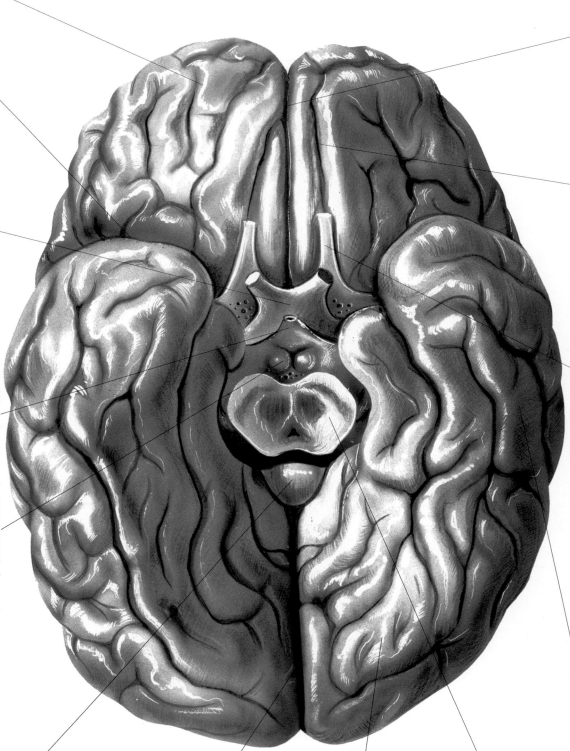

frontal lobe
The frontal lobe forms almost all the anterior part of the cerebrum. Its cortical zone performs the majority of human intellectual activities.

sylvian fissure
A fissure that extends from the base of the cerebrum to its external face and separates the frontal and temporal lobes.

optical chiasm
The optic chiasm is formed by the union of the optic nerves and is the site where fibers from the nasal portion of each retina cross and continue in the opposite optical tract.

stem of the hypophysis
The hypophysis, the gland located at the base of the cerebrum, is united with the hypothalamus through this pedicle.

mamillary bodies
Two hemispherical tubercules formed of gray matter. They contain nervous nuclei corresponding to the hypothalamus.

longitudinal cerebral fissure (anterior part)
A longitudinal fissure located between the two cerebral hemispheres, which extends from the posterior to the occipital pole. The anterior part houses a fibrous wall known as the falx cerebri, so-called because of its sickle shape.

olfactory sulci
Two sulci which run across the inferior faces of the frontal lobes. They contain the olfactory tracts.

olfactory tracts
Two nervous cords that communicate the olfactory sensations captured in the nasal fossas to cerebral centers which interpret them. At one end is the olfactory bulb, located above the cribiform lamina of the ethmoid bone, near the olfactory mucosa of the nasal cavity.

temporal lobe
The temporal lobe is located in the lateral inferior region of each hemisphere. Its cortical zone contains the auditory receptor centers.

splenium of the corpus callosum
The splenium is the rounded, posterior part of the corpus callosum. It is a lamina of white matter that separates the two cerebral hemispheres.

longitudinal cerebral fissure (posterior part)
A longitudinal fissure located between the two cerebral hemispheres and which extends from the posterior to the occipital pole.

occipital lobe
The occipital lobe occupies the posterior part of the cerebral hemispheres. Its cortex contains the sight receptor centers.

cerebral peduncles
Two fibrous columns, united internally, which join the cerebrum and the mesencephalon.

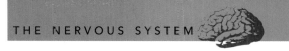

CEREBRUM

▼ SUPERIOR VIEW

CEREBRUM

The largest, central part of the nervous system. It receives all conscious and unconscious impressions arrive and sends out all motor transmissions. In addition, all human intellectual faculties are concentrated in the superficial cerebral cortex. It is located inside the cranial cavity surrounded by the cranial bones.

left cerebral hemisphere
Left portion of the two hemispheres into which the cerebrum is divided. Since the nervous routes which descend from the cerebrum to the rest of the body cross in the zone of the mesencephalon and medulla oblongata, the structures of the left cerebral hemisphere usually dominate over the right in right-handed individuals.

frontal pole
The anterior extreme of the frontal lobe.

right cerebral hemisphere
The cerebrum is divided into right and left lateral hemispheres. The division is external, since in their middle and central portions, the hemispheres are united by the different structures of the base of the cerebrum.

superior frontal sulcus
A sulcus that obliquely crosses the antero-external face of the frontal lobe.

inferior frontal sulcus
A sulcus that runs parallel and inferior to the superior frontal sulcus.

precentral sulcus
A prominent sulcus located between the cerebral gyri, in front of the central sulcus in the frontal lobe.

cerebral gyri
The external surface of the two hemispheres is crossed by wide sulci or fissures that delimit the cerebral gyri. This disposition is caused by the necessity to lodge a great amount of cerebral tissue in a closed cavity such as the skull. They receive the name of the zone in which they are located: superior central circumvolution, median temporal or precentral.

central sulcus (Rolando)
A wide fissure that goes perpendicularly from the middle of the interhemisperic fissure, crosses the external face of the cerebral hemispheres and nearly reaches the sylvian fissure. It separates the frontal and parietal lobes.

intraparietal sulcus
A sulcus which crosses the parietal lobe and delimits its circumvolutions.

superior temporal sulcus
A sulcus that crosses the superior part of the temporal lobe parallel to the sylvian fissure.

postcentral sulcus
A sulcus which separates some of the cerebral gyri of the parietal lobe, following a path parallel to the central sulcus.

longitudinal cerebral fissure
A wide fissure that separates the right and left cerebral hemispheres and extends from the frontal to the occipital poles. In its anterior part, it lodges a fibrous wall, the falx cerebri, which is a prolongation of the meningeal layers which cover the cerebrum.

parieto-occipital sulcus
A sulcus which separates the lobes parietal and occipital lobes. It is also called the external perpendicular fissure.

occipital pole
The posterior extreme of the occipital cavity.

133

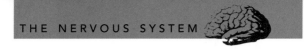

CEREBRUM

▼ EXTERNAL LATERAL VIEW

precentral sulcus
A prominent sulcus between the cerebral gyri, in front of the central sulcus, in the frontal lobe.

central sulcus (Rolando)
A wide fissure that goes perpendicularly from the middle of the interhemisperic fissure, crosses the external face of the cerebral hemispheres and nearly reaches the sylvian fissure. It separates the frontal and parietal lobes.

postcentral sulcus
A sulcus which separates some of the cererbral gyri of the parietal lobe, following a path parallel to the central sulcus.

parieto-occipital sulcus
A sulcus which originates in the posterior third of the interhemispheric sulcus and extends perpendicularly across the superior and external faces of the hemispheres. It separates the parietal and occipital lobes.

transverse occipital sulcus
A sulcus which crosses the external face of the cerebrum vertically, separating the occipital and temporal lobes.

sylvian fissure
A fissure that extends from the base of the cerebrum to its external face and separates the frontal and temporal lobes.

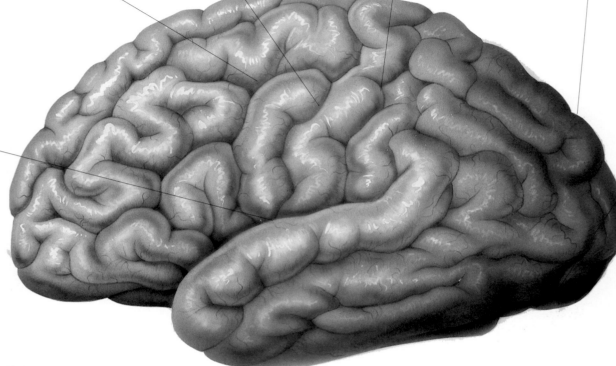

frontal lobe
The frontal lobe forms almost all the anterior part of the cerebrum. It is delimited posteriorly by the central sulcus and inferiorly by the sylvian fissure. Its cortical zone performs the majority of human intellectual activities, with motor activities centered in the gyri anterior to the central sulcus.

parietal lobe
The parietal lobe occupies the posterior part of the cerebral hemispheres, separated from the temporal lobe by the preoccipital sulcus and the occipital lobe by the external perpendicular sulcus. Its cortex contains the visual receptor centers.

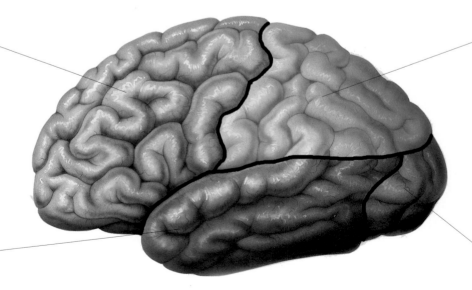

temporal lobe
The temporal lobes are located in the lateral, inferior region of each hemisphere and are separated from the frontal lobe by the sylvian fissure and from the occipital lobe by the preoccipital sulcus. In their posterior superior area, they are prolonged, becoming parietal lobes. Their cortical zones contain the auditory receptor centers.

occipital lobe
Located in the superior external and central part of the cerebral hemispheres, the occipital lobe is separated from the frontal lobe by the central sulcus and the parietal lobe by the external perpendicular fissure. The sensory receptors from all over the body are located in the area posterior to the central sulcus.

CEREBRUM

▼ INTERNAL LATERAL VIEW

septum pelucidum
A medial partition that extends from the corpus callosum to the cerebral trigone separating the lateral ventricles.

corpus callosum
The corpus callosum is abroad, thick band connecting both hemispheres and consisting of a mass of nerve fibers. It consists of anterior, medial and posterior portions.

cerebral trigone
A triangular lamina which originates in the posterior part of the corpus callosum and forms the floor of the septum pelucidum and the roof of the third ventricle.

parietal lobe
Located in the superior external and central part of the cerebral hemispheres. The sensory receptors from all the body are located in the area posterior to the central sulcus.

third ventricle
The third ventricle is a cavity located below the trigone, whose lateral walls are formed by the optical thalamus. It contains cerebrospinal fluid which circulates around the different ventricles and reaches the meningeal spaces, where it protects the cerebrum.

frontal lobe
The frontal lobe comprises almost all the anterior part of the cerebrum. It is delimited posteriorly by the central sulcus and inferiorly by the sylvian fissure. Its cortical zone performs the majority of human intellectual activities, with the motor activities being centered in the gyri anterior to the central sulcus.

occipital lobe
The occipital lobe occupies the posterior part of the cerebral hemispheres. Its cortex contains the visual receptor centers.

135

optical chiasm
The optic chiasm is formed by the union of the optic nerves and is the site at which fibers from the left optical nerve cross fibers obliquely and go to the occipital lobe of the right cerebral hemisphere, and vice versa.

aqueduct of Silvius
A duct that crosses the cerebral peduncles and connects the third ventricle with the fourth ventricle, located inside the mesencephalon and medulla oblongata.

hypophysis
A gland joined to the cerebrum by the hypophysial stem which secretes a series of hormones regulating the operation of the rest of glands in the the organism.

cerebral peduncles
Two fibrous columns, joined internally, which join the cerebrum and the mesencephalon. They are crossed internally by the aqueduct of Silvius.

infundibulum
A funnel-shaped depression located in the floor of the third ventricle, above the hypophysial stem.

commisura grisea
The commisura grisea joins the nuclei of the optical thalami on both sides of the third ventricle. It is also called the interthalamic adherence.

temporal lobe
The temporal lobes are located in the lateral inferior area of each of the hemispheres. The auditory receptor centers are located in their cortical zones.

mamillary bodies
Two hemispherical tubercules, composed of gray matter. They contain nervous nuclei corresponding to the hypothalamus.

CEREBRUM

▼ LONGITUDINAL SECTION

gray matter
The external layer of the cerebral hemispheres, also called the gray cortex. It contains the neuronal bodies, where the nervous signals are drawn up and information integrated.

lateral ventricles
Two cavities located at each side of the cerebrum, which extend from the frontal to the occipital lobe. They contain choroid plexuses which produce the cerebrospinal fluid.

corpus callosum
The corpus callosum is a broad, thick band of white matter connecting both cerebral hemispheres. It serves to join the different parts of the hemispheres.

longitudinal cerebral fissure
A longitudinal fissure, located between the two cerebral hemispheres, which extends from the posterior to the occipital pole.

septum pelucidum
A medial partition that extends perpendicularly from the inferior part of the corpus callosum to the cerebral trigone and separates the lateral ventricles.

third ventricle
The third ventricle is a cavity located below the lateral ventricles with which it is communicated by the foramen of Monro. It contains cerebrospinal fluid which circulates around the different ventricles and reaches the meningeal spaces.

caudate nucleus
One of the basal ganglia of the telencephalon or superior cerebrum. It is a nucleus of gray matter located in the wall of the lateral ventricle that constitutes an important link in the transmission of motor impulses.

white matter
A mass of cerebral tissue located below the gray cortex which surrounds the cerebral nuclei. It contains nervous elements which transmit and conduct signals.

sylvian fissure
A fissure that extends from the base of the cerebrum to its external face and separates the frontal and temporal lobes.

biconvex nucleus
One of the basal ganglia of the telencephalon or superior cerebrum. Like all of them, it constitutes an important link in the transmission of motor impulses.

optical thalamus
Zones of gray matter located on both sides of the third ventricle, which contains groupings of nervous cells that serve as a type of relay station for the nervous pathways that connect with the cerebral cortex.

hippocampus
The hippocampus is located inside the temporal lobe. It forms part of the limbic system and plays a role in memory and direction.

cerebral peduncles
Two fibrous columns, internally united, which join the cerebrum and the mesencephalon. They contain the nerves which enter and leave the cerebrum.

mamillary bodies
Two hemispherical tubercules formed of gray matter. They contain nervous nuclei corresponding to the hypothalamus.

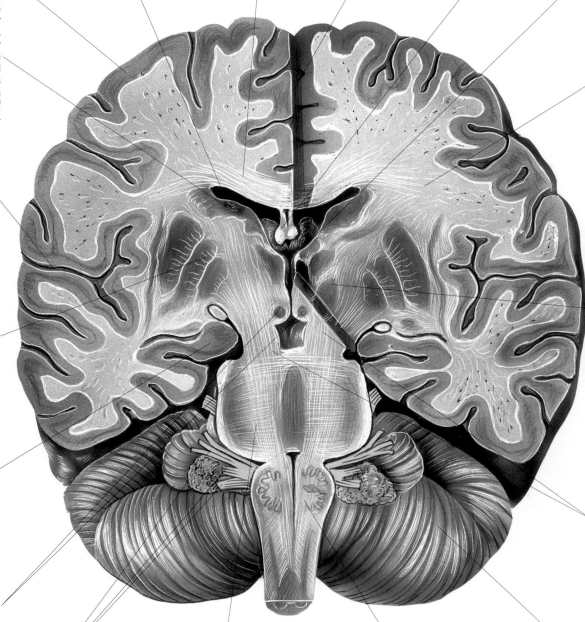

cranial nerves
The seventh (facial), eighth (vestibulocochlear), ninth (glossopharyngeal), tenth (vagus) and twelfth (hypoglossal) nerves which leave the lateral walls of the bulb and the sulcus that separates it from the mesencephalon.

mesencephalon
An eminence located between the medulla oblongata and the base of the cerebrum, with which it is connected by the cerebral peduncles. It contains the nervous pathways that connect the cerebrum with the spinal marrow. Its center contains a cavity called the fourth ventricle.

medulla oblongata
The superior, thicker portion of the spinal marrow. It leaves the cranial cavity through the occipital hole. It contains the nervous pathways that unite the cerebrum with the spinal marrow. It also contains centres which regulate breathing and circulation.

cerebellum
The cerebellum is located below the occipital lobes of the cerebrum and behind the protuberance, in the occipital cerebellar fossa. It is formed by one medial and two lateral lobes. Its main function is to coordinate the movements of many skeletal muscles of the body and it is thus essential for the maintenance of posture, balance and the gati.

CEREBRUM

▼ CROSS-SECTION

septum pelucidum
A medial partition that extends from the corpus callosum to the cerebral trigone and separates the lateral ventricles.

caudate nucleus
One of the basal ganglia of the telencephalon or superior cerebrum. It is a nucleus of gray matter located in the wall of the lateral ventricle that constitutes an important link in the transmission of motor impulses.

longitudinal cerebral fissure
(anterior part)
A longitudinal fissure located midway between the two cerebral hemispheres, which contains a fibrous wall called the falx cerebri.

genu of the corpus callosum
The anterior part of the corpus callosum which consists of a lamina of white matter between both cerebral hemispheres which it connects.

biconvex nucleus
One of the basal ganglia of the telencephalon or superior cerebrum. Like all of them, it constitutes an important link in the transmission of motor impulses.

anterior horns of the lateral ventricles
The anterior zones of the lateral ventricles. The anterior horns extend from the frontal lobe to the occipital lobe. The anterior horns or prolongations are located in the sinus of the frontal lobes. They contain the cerebrospinal fluid.

optical thalamus
Zones of gray matter which contain groupings of nervous cells that serve as a type of relay station of the nervous pathways that connect with the cerebral cortex.

gray matter
The external layer of the cerebral hemispheres also called the gray cortex. Grey matter is also found in the interior of the brain in the different nuclei or specialized nervous groupings. It contains neuronal bodies, where the nervous signals are drawn up and information integrated.

white matter
A mass of cerebral tissue below the gray cortex which surrounds the cerebral nuclei. It contains nervous elements which transmit and conduct signals.

third ventricle
The third ventricle is a cavity located below the lateral ventricles with which it is communicated by the foramen of Monro. It contains the cerebrospinal fluid which circulates around the different ventricles and reaches the meningeal spaces.

central sulcus (Rolando)
A wide fissure that goes perpendicularly from the middle of the interhemisperic fissure and crosses the external face of the cerebral hemispheres.

occipital horns of the lateral ventricles
Posterior parts of the lateral ventricles located in the occipital lobe. The ventricles contain the cerebrospinal fluid.

longitudinal cerebral fissure
(posterior part)
A wide fissure separating the right and left cerebral hemispheres and extends from the frontal to the occipital poles.

splenium of the corpus callosum
The splenium is the rounded, posterior part of the corpus callosum which separates the two cerebral hemispheres.

plexus choroideus
Chord-like formations located in the frontal and occipital horns of the lateral ventricles. They are prolongations of the meninges. Their function is to secrete cerebrospinal fluid.

137

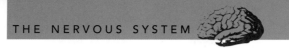

ORIGIN OF THE CRANIAL NERVES

▼ INFERIOR VIEW

oculomotor nerve (III pair)
The nerve which transmits motor orders to all the ocular musculature. It originates in the cerebral peduncles and enters the ocular cavity through the orbital fissure.

trochlear nerve (IV pair)
A nerve with a long intracranial passage reaching from its origin in the lateral zone of the cerebral peduncles to its termination in the ocular cavity. It ends in the greater oblique muscle of the eye, whose operations it controls.

trigeminal nerve (even V)
A nerve that receives the sensations of the face, orbit, oral cavity and the nasal fossas and transmits the motor orders to the chewing muscles. It originates in the mesencephalon, where a knot, called the ganglion of Gasser, forms and from which its three branches, the ophthalmic, maxillary and mandibular, emerge.

facial nerve (VII pair)
A sensory-motor nerve that originates in the pontomedullary sulcus. It has two branches, the proper facial nerve and the sensory intermediate nerve. It goes laterally towards the internal acoustic duct, crosses the petrous bone and emits tympanic, auricular, lingual, temporal, facial and cervical branches and others to the parotid gland.

intermediate nerve
A sensory branch of the facial nerve that innervates the lingual, sublingual and submaxillary glands.

vestibulocochlear nerve (VIII pair)
A sensory nerve that originates in the pontomedullary sulcus and penetrates the internal acoustic duct. It transmits auditory sensations and other postural transmissions that help to maintain the balance.

glossopharyngeal nerve (IX pair)
A sensory-motor muscle which originates in the medulla oblongata. It exits the skull through the posterior jugular foramen and emits nervous terminations, some of which join with the facial nerve and others which go to the tympanic cavity, penetrating the petrous bone, towards the carotid zone, lingual and pharyngeal and innervating some of the pharyngeal muscles.

optical nerve (II pair)
The optical nerves are two nervous structures that transport the visual sensations collected in the terminations of the ocular retina to the interior of the brain.

olfactory tracts
Two nervous cords that communicate the olfactory sensations captured in the nasal fossas to cerebral centers which interpret them. At one end is the olfactory bulb, located above the cribiform lamina of the ethmoid bone, near the olfactory mucosa of the nasal cavity, to which the nerve fibers that constitute the two olfactory nerves or first cranial pair are united.

vagus nerve (X pair)
A sensory-motor nerve that originates in the medulla oblongata; it leaves the cranial cavity by the jugular foramen and descends through the neck and thorax to the abdomen, emitting many nervous branches in its path.

spinal marrow
A long, almost cylindrical cord which is the continuation of the medulla oblongata and descends down the back, contained by the spinal column or vertebral column. It contains the spinal nerves which go to all parts of the body.

abducens nerve (VI pair)
An exclusively motor nerve, that originates in the sulcus separating the mesencephalon from the medulla oblongata. It goes to the ocular cavity from where it emits branches to the rectus lateralis muscle of the eye.

mesencephalon
An eminence located between the medulla oblongata and the base of the cerebrum, with which it is connected by the cerebral peduncles. It contains the nervous pathways that communicate the cerebrum with the spinal marrow.

medulla oblongata
The superior, thicker portion of the spinal marrow from which contains the internal nuclei which originate various cranial nerves. It also contains the centers which regulate breathing and circulation.

hypoglossal nerve (XII pair)
Motor muscle that originates in the lateral zone of the bulb and innervates a large part of the lingual musculature.

cerebellum
An intracranial organ located below the occipital lobes of the brain, behind the mesencephalon and over the medulla oblongata. It serves to coordinate the movements of many of the skeletal muscles of the body.

spinal or accessory nerve (XI pair)
A sensory-motor nerve formed by the union of several nervous branches that originate in the medulla oblongata and the spinal marrow. It emits ramifications to the soft palate, larynx, pharynx and to the trapezius and sternocleidomastoid muscles and has a branch which connects with the vagus nerve.

CEREBELLUM

▼ POSTERIOR VIEW

superior vermis
A swelling that constitutes the central lobe of the cerebellum, located between the two cerebellar hemispheres.

cerebellar hemispheres
The two lateral lobes of the cerebellum. Their surface is marked by a series of parallel sulci.

CEREBELLUM
An intracranial organ located below the occipital lobes of the brain and behind the protuberance lodged in the occipital cerebellar fossa. It is composed of two lateral lobes and a medial lobe. Its main function is to coordinate the movements of the skeletal muscles of the body, and it is vital for functions such as maintenance of the bodily position and the balance.

▼ ANTERIOR VIEW

fourth ventricle
The posterior wall of the fourth ventricle is formed by a lamina called the tectorial membrane, which is attached to the anterior face of the cerebellum.

cerebellar peduncles
Structures that connect the cerebellum with the mesencephalon. Bunches of nerve fibers that connect both parts of the nervous system run through them. There are superior, medial and inferior peduncles.

fissures of the cerebellum
Multiple fissures that furrow the surface of the cerebellum. They penetrate more or less deeply in the cerebellar tissue.

inferior vermis
The superior vermis extends across the inferior face of the cerebellum, maintaining a constant shape.

horizontal fissure
A sulcus dividing the anterior face of the cerebellar hemispheres into superior and inferior parts

▼ HORIZONTAL CROSS-SECTION

fourth ventricle
A cavity located between the mesencephalon and the cerebellum. Where it runs between the two cerebellar peduncles, it is also called the rhomboid fossa. It contains the cerebrospinal fluid.

vermis
The central lobe of the cerebellum which runs from front to back. It is composed of various zones known as the lingual, pyramid, uvula, nodule and tuber vermis.

cerebellar cortex
A thin layer of gray matter, which is very rich in neurons, and which comprises the external part of the cerebellum, covering all the cerebellar sulci and elevations.

cerebellar nuclei
Structures composed of gray matter located in the sinus of the cerebellar tissue. They are classified as dentate, globose, emboliform and fastigial nuclei. They receive nerve fibers from the cerebellar cortex and emit others which go to other parts of the nervous system.

medullar substance
A white substance, formed by nervous prolongations of the neurons located in the gray matter of the cortex.

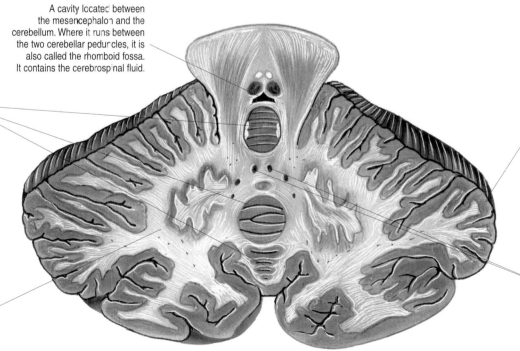

MEDULLA OBLONGATA AND MESENCEPHALON

▼ ANTERIOR VIEW

cerebral peduncles
Two fibrous columns, united internally, which connect the cerebrum and the mesencephalon.

oculomotor nerve (III pair)
The nerve which transmits motor orders to all the ocular musculature.

olfactory tract
A nervous termination that communicates the olfactory sensations captured in the nasal fossas to the cerebral centers which interpret them.

optical chiasm
The optic chiasm is formed by the union of the optic nerves and is the site at which fibers from the left optical nerve cross fibers obliquely and go to the occipital lobe of the right cerebral hemisphere, and vice versa.

stem of the hypophysis
The hypophysis, the gland located in the base of the cerebrum in the cavity of the sphenoid bone called the sella turcica, is united with the hypothalamus through this pedicle.

mamillary bodies
Two hemispherical tubercules, formed of gray matter. They contain nervous nuclei corresponding to the hypothalamus.

trochlear nerve (IV pair)
A motor nerve that goes to the orbit and is responsible for the mobility of the greater oblique muscle of the eye.

trigeminal nerve (V pair)
A mixed nerve (motor and sensory) that receives transmissions from the face, orbit, oral cavity and nasal fossas, and transmits motor orders to the chewing muscles. It has three branches: the ophthalmic, maxilla and mandibular nerves.

facial nerve (VII pair)
A sensory-motor nerve with two branches: the proper facial nerve and the sensory nerve which is the intermediate nerve that innvervates the lingual, sublingual and submaxillary glands.

mesencephalon
The mesencephalon is located between the medulla oblongata and the base of the brain. It connects with the brain through the cerebral peduncles. It contains the nerves that communicate the brain with the spinal marrow.

vestibulocochlear nerve (VIII pair)
A sensory nerve which transmits auditory sensations to the cochlear area and collects sensations from the vestibular area of the ear, helping to maintain the balance.

abducens nerve (VI pair)
A motor nerve that goes to the ocular cavity and innervates the external rectum muscle of the eye.

glossopharyngeal nerve (IX pair)
A sensory-motor nerve which emits nervous terminations, some of which are united with the facial nerve and others which go to the tympanic cavity and the carotid, lingual and pharyngeal areas.

pontomedullary sulcus
A sulcus that separates the mesencephalon from the medulla oblongata.

medulla oblongata
The superior, thicker portion of the spinal marrow. Its interior contains the nerves which join the brain with the spinal marrow. It also contains the centers which regulate breathing and circulation.

anterior medial sulcus
A sulcus that crosses the anterior face of the medulla oblongata and extends down the spinal marrow in the same direction.

hypoglossal nerve (XII pair)
A motor muscle that innervates a large part of the lingual musculature.

spinal or accessory nerve (XI pair)
A sensory-motor nerve that emits ramifications to the villi of the palate, larynx, pharynx and to the trapezius and sternocleidomastoid muscles.

cerebellum
An intracranial organ located below the occipital lobes of the brain, behind the mesencephalon and over the medulla oblongata. It serves to coordinate the movements of many skeletal muscles of the body, making it essential for maintaining the body's posture and balance.

vagus nerve (X pair)
A sensory-motor nerve that leaves the cranial cavity and descends through the neck and thorax to the abdomen, emitting many nervous branches in its path.

MEDULLA OBLONGATA AND MESENCEPHALON

▼ INTERNAL VIEW

commisura grisea
The commisura grisea joins the nuclei of the optical thalami on both sides of the third ventricle It is also called the interthalamic adherence.

third ventricle
The third ventricle is a cavity located below the lateral ventricles with which it is communicated by the foramen of Monro. It contains cerebrospinal fluid which circulates around the different ventricles and reaches the meningeal spaces.

corpus callosum
The corpus callosum is a broad, thick band of white matter, located between the cerebral hemispheres, which connects the different areas of the hemispheres.

aqueduct of Silvius
A duct that crosses the cerebral peduncles and communicates the third ventricle with the fourth ventricle, allowing the cerebrospinal fluid to flow between the two spaces.

mamillary tubercules
Two half-moon shaped papilla formed of gray matter. They contain nervous nuclei corresponding to the hypothalamus.

pineal gland or epiphysis
A gland located in the posterior wall of the third ventricle that secretes a hormone known as melatonin.

141

optical chiasm
The optic chiasm is formed by the union of the optic nerves and is the site at which fibers from the left optical nerve cross obliquely and go to the occipital lobe of the right cerebral hemisphere, and vice versa.

quadrigeminal bodies
Four tubercules, located in the posterior face of the mesencephalon, which contain nervous nuclei that take part in the transmission of visual and auditory sensations.

hypophysis
A gland joined to the cerebrum by the hypophysial stem which secretes a series of hormones regulating the operation of the rest of glands of the organism.

cerebellum
An intracranial organ located behind the mesencephalon and above the medulla oblongata. It is composed of two lateral lobes and a medial lobe. Its main function is to coordinate the movements of the skeletal muscles and it is vital for the maintenance of posture and balance.

mesencephalon
An eminence located between the medulla oblongata and the cerebral peduncles. It contains the nervous pathways that communicate the cerebrum with the spinal marrow.

medulla oblongata
The superior, thicker portion of the spinal marrow. Its interior contains the nerves which connect the brain with the spinal marrow. It also contains the centers which regulate breathing and circulation.

fourth ventricle
A cavity located between the mesencephalon and the cerebellum, in which the aqueduct of Silvius, coming from the medial ventricle, ends. It continues inferiorly as the central medullary duct. The cerebrospinal fluid circulates in the fourth ventricle and goes to the subarchnoid space of the meninges through orifices located in the lateral face of the fourth ventricle.

SPINAL MARROW

SPINAL MEDULLA

Part of the central nervous system which leaves the cranial cavity and crosses the trunk vertically through the vertebral column. It is almost cylindrical and sends out nerves which go to all parts of the body.

medulla oblongata

The superior, thicker portion of the spinal marrow. It contains the nervous pathways that unite the brain with the spinal marrow. It also contains the vital centers which regulate breathing and circulation.

cerebellum

An intracranial organ located behind the mesencephalon and above the medulla oblongata. It is composed of two lateral lobes and one medial lobe. Its main function is to coordinate the movements of the skeletal muscles and it is vital for the maintenance of balance.

posterior medial sulcus

A sulcus which crosses the posterior face of the spinal marrow vertically, beginning in the medulla oblongata and terminating in the sacral area.

spinal nerves

Lateral branches which are emitted by the spinal marrow throughout all its trajectory. They leave the vertebral column through the intervertebral foramens. The spinal nerves innervate all the areas of the body. The nerves are divided into anterior and posterior branches immediately after leaving the vertebral column. There are 31 pairs of spinal nerves: eight cervical, twelve dorsal, five lumbar, five sacral and one coccygeal.

phrenic nerve

A nerve that originates in the cervical plexus and descends down the neck and thorax to reach the diaphragm which it innervates.

cervical plexus

The cervical plexus is formed by the union of the anterior branches of the four first cervical spinal nerves, which give rise to a series of nervous ramifications that innervate all the structures of the neck.

brachial plexus

The brachial plexus is formed by the union of the anterior branches of the fifth, sixth, seventh and eighth cervical spinal nerves as well as the first thoracic nerve, giving rise to three thick nervous trunks, from which the nerves come that innervate the superior extremity.

intercostal nerve

All the spinal nerves divide into anterior and posterior branches. The anterior branches of the dorsal spinal nerves follow a path parallel to the ribs in the intercostal spaces and are thus called the intercostal nerves. They innervate the intercostal muscles and the last intercostal nerves also innervate the muscles of the abdominal wall.

vertebral pedicles

The two lateral walls of the vertebral foramen which contain the intervertebral foramens between the vertebrae, through which the spinal nerves pass.

intervertebral foramens

Orifices delimited by the vertebral pedicles, through which the different somatic nerves emerge from the vertebral column.

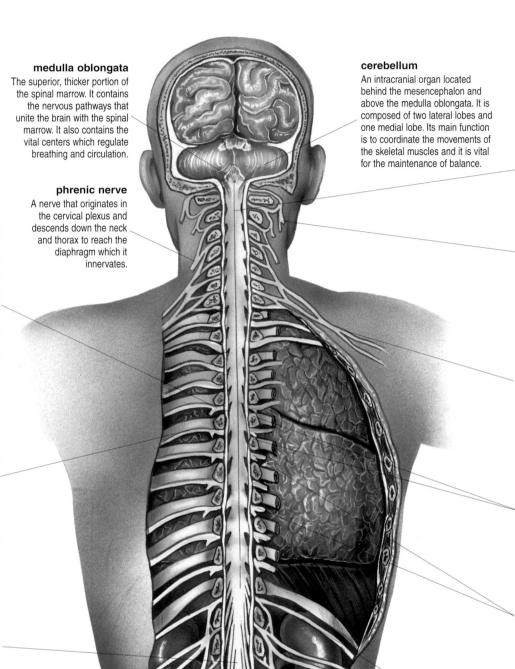

medullar cone

The conical ending of the spinal cord. It continues as a thin, fiber stripe

subcostal nerve

The last of the intercostal nerves. It does not pass between two ribs, but under the last rib. It follows a path similar to the intercostal nerves and descends to the gluteal region.

cauda equina

A bunch of nervous cords that descend vertically and obliquely from the medullar cone. It is formed by the nervous roots of the three last lumbar spinal nerves and the sacral and coccygeal nerves.

lumbosacral plexus

A plexus formed by the union of the anterior branches of the lumbar spinal nerves and those of the first three sacral nerves. It innervates all the inferior extremity, the inferior abdominal area and the genital area.

spinal dura mater

The spinal marrow is covered, as are other parts of the central nervous system, by the meninges, which are three membranous layers, called, externally to internally, the dura mater, arachnoid membrane and pia mater. The dura mater extends downwards, beyond the spinal marrow, forming a pouch which reaches the second sacral vertebra.

terminal filum

A thin, rudimentary prolongation formed by the marrow as a continuation of the medullar cone. It reaches the coccyx where it inserts itself.

MENINGES

superior longitudinal sinus
A venous duct in the dura mater which crosses the cerebral interhemispheric area from front to back above the sickle of the brain.

scalp
The layer of skin that covers the skull and is generally covered by hair.

subcutaneous cellular tissue
The deepest layer of the skin. It is composed of fatty tissue which serves to cushion and protect the deeper structures.

skull
The bony structure that surrounds the brain. It is formed of two laminae called diploe that surround a central area of spongy tissue

dura mater
The external, thickest layer of the meninges. It is attached to the periosteum or internal layer of the cranial bone. It has a fibrous structure and it acts to protect the cerebral structures and maintain them in position.

MENINGES
Layers that cover the central nervous system from the brain to the spinal marrow. There are three superimposed layers which are the dura mater, arachnoid membrane and pia mater.

meningeal arteries and veins
A dense network of arteries and veins that run through the meninges. The arteries come from three main branches: the anterior, median and posterior meningeal arteries. The veins go to the venous sinuses surrounding the brain.

subdural space
The space between the dura mater and the arachnoid membrane. It is a very narrow space, almost non-existent in parts because the two meningeal layers are attached to each other over a large part of their surface. The meningeal veins, arteries and nerves pass through the space.

143

arachnoid membrane
A thin, fibrous membrane which is attached to the internal face of the dura mater and has a similar extension.

subarchnoid space
A somewhat wider space between the arachnoid membrane and the pia mater. The two membranes are not attached to each other, thus making the subarachnoid space wider. It contains cerebrospinal fluid, which serves to protect the brain from any impact.

cerebral cortex
The superficial part of the brain, located immediately below the piameter, is formed of gray matter containing many neurons with specific functions. The cerebral cortex contains the mechanisms responsible for memory, the elaboration of thought, manual skills and speech.

pia mater
The internal layer of the meninges, which is attached to the external surface of the brain, cerebellum and spinal marrow, covering all the irregularities produced on the surface of these organs.

falx cerebri
A fibrous prolongation of the dura mater, which is contained in the longitudinal fissure and separates the cerebral hemispheres.

white matter
A mass of cerebral tissue located below the cerebral cortex which surrounds the different cerebral nuclei. It mainly contains nervous elements of transmission and conduction.

LUMBOSACRAL PLEXUS

LUMBOSACRAL PLEXUS

A plexus formed by the union of the anterior branches of the lumbar spinal nerves and those of the first three sacral nerves. It innervates all the inferior extremity, the inferior abdominal area and the genital area.

subcostal nerve
The last of the intercostal nerves. It does not pass between two ribs but under the last rib. It does not belong to the lumbosacral plexus but follows a path similar to the intercostal nerves, descending to the gluteal region.

iliohypogastric nerve
Also known as the major abdominal-genital nerve. It originates in the anterior branch of the first lumbar spinal nerve and gives rise to branches going to the gluteal region, or to others that innervate the inferior zone of the abdominal wall and others that descend through the inguinal canal to the genital area and the superior part of the thigh.

ilioinguinal nerve
Also known as the minor abdominal genital nerve. It originates in anterior branch of the first lumbar spinal nerve. It follows a similar path to the iliohypogastric nerve and collaborates with it in the innervation of the inferior abdomen and the genital area.

femoral cutaneous nerve
A nerve which originates between the anterior branches of the second and third lumbar spinal nerves. It follows a descending path and leaves the abdominal cavity passing below the crural arch. In the inferior extremity, it divides into an anterior or femoral branch and a posterior or gluteal branch, which innervate the superficial cutaneous zones of these regions.

genitofemoral nerve
A nerve that originates in the anterior branch of the second lumbar spinal nerve. It bifurcates into the external or the femoral branch which crosses the crural arch and goes to the superior part of the thigh, and the internal or genital branch which goes through the inguinal canal to the scrotum in men and the labia majora in women.

femoral nerve
A thick nerve that originates in the junction between the anterior branches of the second, third and fourth lumbar spinal nerves. It goes to the inferior extremity through the crural arch and divides into four branches: the external and internal musculocutaneous nerve, the quadriceps nerve and the internal saphenous nerve.

sympathetic trunk
A part of the autonomic nervous system formed by a series of joined nervous ganglia that run down the vertebral column from the thoracic cavity to the abdominal cavity, emitting nervous prolongations to the organs of these regions.

lumbosacral trunk
The result of the union of the anterior branches of the fourth and fifth lumbar spinal nerves. It descends to unite with the anterior branches of the first spinal sacral nerves and the sciatic nerve.

pudendal nerves
Nerves originating in the junction between the anterior branches of the second and third spinal sacral nerves. They descend to the genital zone, where they send branches to the perineum and to the penis in men and the clitoris in women.

anal nerve
A nerve that originates at the meeting point between the anterior branches of the third and fourth spinal sacral nerves, following a parallel path to the pudendal nerve to finally reach the anal region.

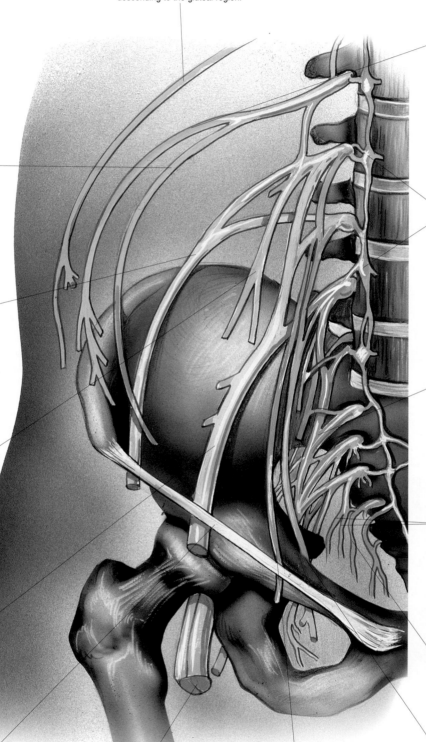

crural arch
A thin, fibrous ligament that extends obliquely from the anterosuperior iliac spine to the pubis. It marks the limit between the pelvic and femoral regions. The vessels and nerves that go to the inferior extremity pass through the crural arch.

sciatic nerve
The largest nerve in the body. It originates in the union of the lumbosacral trunk with the anterior branches of the first spinal sacral nerves. It leaves the pelvis through the greater sciatic notch and, after passing behind the hip joint, it descends the posterior part of the thigh, dividing into two branches at the height of the popliteal fossa of the knee: the external and internal sciatic popliteal nerves.

accessory obturator nerve
A nerve which follows a parallel path to the obturator nerve. It is not present in all people.

obturator nerve
A nerve formed by the union of the anterior branches of the second, third and fourth lumbar spinal nerves. It descends to the pelvic cavity where it divides into different branches going to the adductor muscles of the thigh.

BRACHIAL PLEXUS

BRACHIAL PLEXUS

The brachial plexus is formed by the union of the anterior branches of the fifth, sixth, seventh and eighth cervical spinal nerves and the first thoracic nerve. It gives rise to the superior, middle and inferior trunks which are the origin of all the nerves going to the superior extremity.

musculocutaneous nerve

A nerve which originates in the superior trunk of the brachial plexus. It crosses the external part of the arm and the forearm, giving motor branches to the muscles of the anterior face of the arm and sensory branches to the skin of the forearm, where some of its terminal branches reach the wrist (lateral antebrachial cutaneous nerve).

radial nerve

A nerve where the middle trunk of the brachial plexus originates. It goes to the axilla, passes the posterior face of the arm, crossing it behind the humerus and divides into an anterior or sensory branch and a posterior or muscular branch at the elbow. In its path down the arm, the radial nerve sends muscular branches to the triceps and other muscles of the zone as well as sensory branches to the skin (posterior antebrachial cutaneous nerve).

superior trunk

Formed by the union of the anterior branches of the fifth and sixth spinal nerves, with a small branch from the fourth. The superior trunk is the origin of the musculocutaneous nerve and part of the median nerve. A posterior branch joins the middle trunk to give rise to the radial nerve.

middle trunk

The middle trunk originates exclusively in the anterior root of the seventh cervical spinal nerve. Near the axilla, it gives off a branch that is united to the prolongation of the superior trunk. The middle trunk has a posterior branch where the radial nerve originates.

axillary (circumflex) nerve

A nerve which originates in the middle trunk of the brachial plexus, from which the radial nerve emerges. It separates from the path of the radial nerve and, after crossing below the shoulder joint, it terminates in that area, sending out articular branches for the shoulder joint, motor branches to the deltoid muscle and others, and sensory branches to the skin of the shoulder.

inferior trunk

Its origin is in the union of the anterior branches of the eighth cervical spinal nerve and the first thoracic. It gives rise to the ulnar, cutaneous brachial medial and cutaneous antebrachial medial nerves. It also has a branch that contributes to forming the median nerve and another posterior branch where the radial nerve originates.

long chest nerve

A nerve where the small posterior branches of the fifth, sixth and seventh spinal nerves originates and descends vertically towards the lateral wall of the thorax to innervate the anterior serratus muscle.

pectoral nerves

Anterior collateral branches of the brachial plexus that innervate the bigger and smaller pectoral muscles.

subscapular nerves

Posterior collateral branches of the brachial plexus which innervate the subscapular and teres major muscles.

median nerve

The median nerve originates at the height of the axilla through the union of branches of the superior and inferior trunks. It descends the internal border of the anterior face of the arm, crosses the elbow joint and continues in the central zone of the anterior face of the forearm, crossing the wrist and terminating in the palm of the hand. It gives off the majority of its muscular and sensory branches in the forearm and the hand, sending out only a few branches to the humerus and the elbow joint in the arm.

ulnar nerve

A nerve which originates in the inferior trunk of the brachial plexus, crosses the internal border of the arm and the elbow and extends down the forearm to the hand. In its trajectory down the arm, it emits no nervous branch, and in the forearm, it gives rise to nervous branches that go to the elbow joint and the muscles of its internal face.

medial antebrachial cutaneous nerve

A nerve which originates in the same trunk as the ulnar nerve and descends down the arm parallel to it. After crossing the elbow, it is distributed in multiple sensory branches in the internal face of the forearm.

medial brachial cutaneous nerve

A nerve which originates in the same trunk as the ulnar nerve, above the medial antebrachial cutaneous nerve, descending with these two nerves to terminate in a series of sensory branches that go to the internal and posterior areas of the skin of the arm.

145

ARM

axillary (circumflex) nerve

A nerve which originates in the middle trunk of the brachial plexus, from which the radial nerve emerges. It separates from the path of the radial nerve and, after crossing below the shoulder joint, it terminates, sending out articular branches for the shoulder joint, motor branches to the deltoid muscle and sensory branches to the skin of the shoulder.

musculocutaneous nerve

A nerve which originates in the brachial plexus formed from the union of the fifth and sixth spinal nerves. It crosses the external part of the arm and forearm, giving motor branches to the muscles of the anterior face of the arm and sensory branches to the skin of the forearm, where some of its terminal branches reach the wrist (lateral antebrachial cutaneous nerve).

posterior antebrachial cutaneous nerve

A branch of the radial nerve that acts in the posterior superficial areas of the arm and forearm.

radial nerve

A nerve which originates in the brachial plexus from the union of the sixth, seventh and eighth cervical nerves and the first thoracic nerve. It goes to the axilla, passes the posterior face of the arm, crossing it behind the humerus, and divides into an anterior or sensory branch and a posterior or muscular branch at the elbow. In its path down the arm, the radial nerve sends muscular branches to the triceps and other muscles of the zone as well as sensory branches to the skin (posterior antebrachial cutaneous nerve).

lateral antebrachial cutaneous nerve

A prolongation of the musculocutaneous nerve in the forearm, which it reaches by the anterior face of the elbow joint. Here it sends multiple sensory terminations to the skin of the forearm and the wrist.

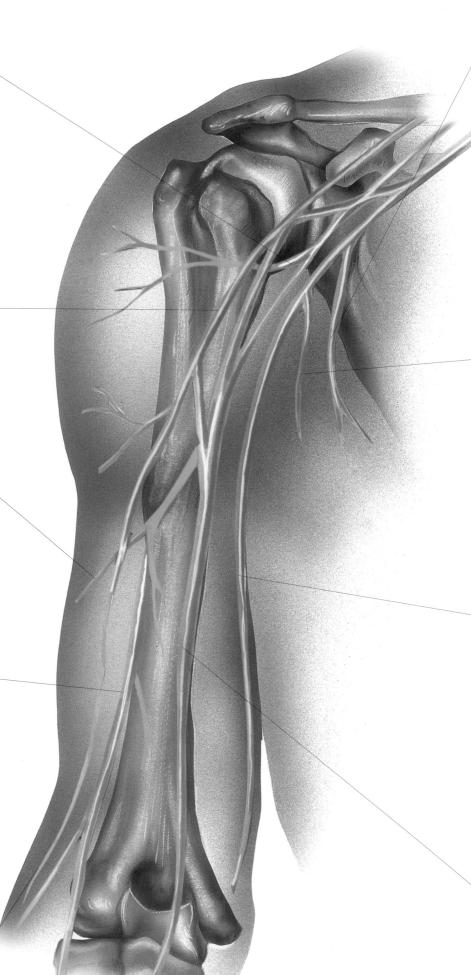

medial brachial cutaneous nerve

A nerve which originates in the same trunk as the ulnar nerve, above the medial antebrachial cutaneous nerve, descending with these two nerves to terminate in a series of branches that go to the internal posterior area of the skin of the arm.

medial antebrachial cutaneous nerve

A nerve which originates in the same trunk as the ulnar nerve and descends down the arm parallel to it. After crossing the elbow, it is distributed in multiple sensory branches in the internal face of the forearm.

ulnar nerve

A nerve which originates in the brachial plexus from the union of the eighth cervical and first thoracic nerve. It crosses the internal border of the arm and the elbow and extends down the forearm to the hand. In its trajectory down the arm, it gives off no nervous branch.

median nerve

A nerve which originates in the brachial plexus from the union of the sixth, seventh and eighth cervical and first thoracic spinal nerves. It descends the anterior face of the arm, crosses the elbow joint and continues in the central zone of the anterior face of the forearm, crossing the wrist and terminating in the palm of the hand. It gives off the majority of its muscular and sensory branches in the forearm and the hand, sending out only a few branches to the humerus and the elbow joint in the arm.

FOREARM AND HAND

lateral antebrachial cutaneous nerve

A prolongation of the musculocutaneous nerve in the forearm, which it reaches through the anterior face of the elbow joint. Once in this zone, it branches into multiple sensory terminations that reach the skin of the forearm and the wrist.

radial nerve

After crossing the posterior face of the arm, where it has muscular and sensory branches, the radial nerve reaches the elbow, which it crosses in its external part to reach the forearm, where it divides into two branches, one superficial sensory branch and another deep muscular branch.

superficial terminal branch of the radial nerve

The radial nerve divides into two muscular branches when it reaches the forearm. The superficial terminal branch crosses the posteroexternal part of the forearm. It descends to innervate the external dorsal face of the hand.

deep terminal branch of the radial nerve

The radial nerve divides into two muscular branches when it reaches the forearm. The deep terminal branch sends nervous branches to the muscles of the posterior face of the forearm.

common palmar digital nerves

When it reaches the palm of the hand, the median nerve gives off different branches that innervate the muscles of the thenar eminence and the palm of the hand. It has branches to the first, second, third and fourth.

proper palmar digital nerves

The common palmar digital nerves emit nervous branches that innervate the first, second and third fingers and part of the fourth.

median nerve

The median nerve descends the arm, crosses the elbow joint and continues in the central zone of the anterior face of the forearm, crossing the wrist and terminating in the palm of the hand. It gives off the majority of its muscular and sensory branches in the forearm, innervating the muscles of the forearm's anterior face. After crossing the wrist, it divides into several branches that go to the fingers.

anterior interosseal nerve

Together with the branches for the muscles of the anterior face of the forearm, the median nerve gives off a branch that goes to the interosseal space located between the ulna and radius and innervates the muscles of the space.

ulnar nerve

The ulnar nerve crosses the internal border of the arm and the elbow behind the epitrochlea and extends down the forearm to the hand. In its trajectory down the arm, it gives off no nervous branch. In the forearm, it gives rise to nervous branches that go to the elbow joint and the muscles of the internal face of the forearm and, after crossing the wrist, sends a sensory branch to the dorsal face of the hand and gives off deep and superficial terminal branches.

deep terminal branch of the ulnar nerve

The ulnar nerve bifurcates into two branches when it reaches the palm of the hand. The deep branch goes transversally to the thumb, innervating some muscles of the little finger, the thumb and the interosseal spaces.

superficial terminal branch of the ulnar nerve

A more internal branch which goes to the hypothenar eminence where it gives off digital branches that go to the fourth and fifth fingers.

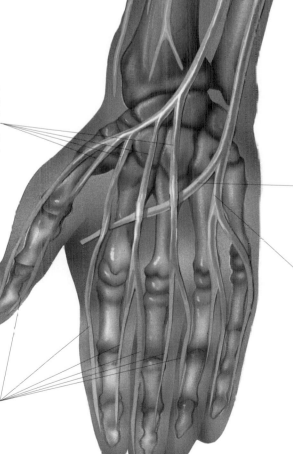

147

LEG AND FOOT

sciatic nerve

The largest nerve in the body. It originates in the union of the lumbosacral trunk with the anterior branches of the first spinal sacral nerves. It leaves the pelvis through the greater sciatic notch and, after passing behind the hip joint, it descends the posterior part of the thigh, dividing into two branches at the height of the popliteal fossa of the knee: the external and internal sciatic popliteal nerves.

external sciatic popliteal nerve

The external branch of the two into which the sciatic nerve is divided. It goes outwards behind the tibiofibular joint and borders the head of the fibula, passing to the anteroexternal face of the leg. It bifurcates into two terminal branches: the superficial and deep fibular nerves. In its short passage, the external sciatic popliteal nerve has articular branches that go to the knee joint and cutaneous branches to the skin of the region.

superficial fibular nerve

The external bifurcation of the external sciatic popliteal nerve, which crosses the exterior part of the leg vertically, following a parallel path to the fibula and sending muscular branches to the musculature and cutaneous branches of the region. Near the ankle joint, the superficial fibular nerve bifurcates into the internal and external cutaneous dorsal nerves.

anterior deep tibial nerve

The internal bifurcation of the external sciatic popliteal nerve. It crosses the anterior face of the leg vertically, in front of the tibia, and reaches the ankle joint to pass to the dorsal zone of the foot. It has nervous branches to the musculature of the anterior face of the leg, the ankle joint and the internal dorsal surface of the foot.

external dorsal cutaneous nerve

One of the terminal branches of the superficial fibular nerve. It runs down the external face of the dorsal part of the foot and innervates the third, fourth and fifth toes.

internal dorsal cutaneous nerve

One of the terminal branches of the superficial fibular nerve that crosses the internal face of the dorsal part of the foot, innervating the first and second toes.

tibial nerve

The external bifurcation of the sciatic nerve. It continues the posterior path of the sciatic nerve and descends the leg behind the tibia. It has branches to the knee joint, the musculature of the dorsal part of the leg and the skin of the region. When it reaches the ankle joint, it passes behind the medial malleolus and goes to the sole of the foot, where it divides into the internal and external plantar nerves. It also has a branch to the skin of the heel called the internal calcaneal nerve.

medial sural cutaneous nerve

A nerve originating in the tibial nerve and crossing the back of the leg superficially to reach the ankle, which it passes to reach the external border of the foot.

internal saphenous nerve

One of the branches of the femoral nerve which divides in the superior part of the thigh. It crosses the thigh, the knee and the leg internally and, crossing the ankle joint in front of the medial malleolus, terminates in the internal border of the foot. It has nervous branches to the skin of the internal face of the thigh, the knee, the leg and the foot.

external plantar nerve

After crossing behind the lateral malleolus, the tibial nerve reaches the sole of the foot and gives off an external branch that innervates the musculature and skin of the area.

internal plantar nerve

When it reaches the sole of the foot, the tibial nerve gives off an internal branch which innervates the musculature and skin of the area and has digital nervous branches which go to the first, second and third toes.

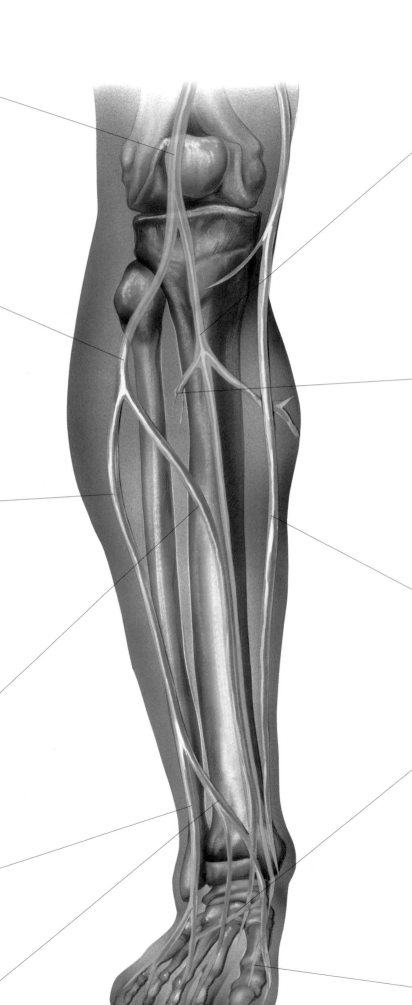

THIGH

femoral cutaneous nerve
A nerve which originates in the first roots of the lumbosacral plexus. In the inferior extremity, it divides into an anterior or femoral branch and a posterior or gluteal branch, which innervate the superficial cutaneous zones of these regions.

femoral nerve
A thick nerve that originates in the lumbosacral plexus. It goes to the inferior extremity through the crural arch and divides into four branches: the external and internal musculocutaneous nerve, the quadriceps nerve and the internal saphenous nerve.

obturator nerve
A nerve formed by the union of the anterior branches of the second, third and fourth lumbar spinal nerves. It descends to the pelvic cavity where it divides into different branches that go to the adductor muscles of the thigh.

external cutaneous branch of femoral nerve
A branch of the femoral nerve that innervates the sartorius muscle and gives off prolongations to the superficial area of the anterior face of the thigh, innervating the skin.

vastus lateralis nerve
A branch of the quadriceps nerve that innervates the external portion of this muscle group, the vastus lateralis muscle.

rectus femoris nerve
A branch of the quadriceps nerve that innervates the anterior portion of this muscle group, the rectus femoris muscle.

crural arch
A thin, fibrous ligament that extends obliquely from the anterosuperior iliac spine to the pubis. It marks the limit between the pelvic and femoral regions. The vessels and nerves that go to the inferior extremity pass through the crural arch.

sciatic nerve
The largest nerve in the body. After its origin in the lumbosacral plexus it leaves the pelvis through the greater sciatic notch and passes behind the hip joint, arriving at the posterior part of the thigh by the gluteal region. It crosses the thigh from top to bottom, emitting branches for the muscles of that zone and, when it arrives at the height of the popliteal fossa of the knee, it is divided into two branches: the external sciatic popliteal nerve and the tibial nerve.

internal cutaneous branch of the femoral nerve
A branch of the femoral nerve that innervates some muscles of the superointernal part of the thigh and the skin that covers it.

internal saphenous nerve
One of the branches of the femoral nerve which crosses the internal face of the thigh and reaches the knee, sending branches to the skin and joints of this area. It then divides into patellar and tibial branches.

branch of femoral nerve to the quadriceps
A branch of the femoral nerve located in the central part of the anterior face of the thigh. It innervates the quadriceps muscle, emitting nervous prolongations to each of its four parts.

vastus medialis nerve
A branch of the quadriceps nerve that innervates the internal portion this muscle group, the vastus medialis muscle.

crural nerve
A branch of the quadriceps nerve that innervates the middle portion of this muscle group, the vastus intermedius muscle.

149

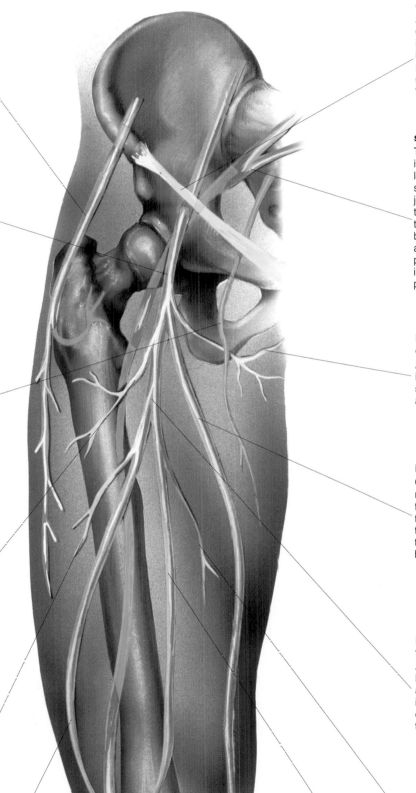

sciatic nerve

VISION: THE EYE

pupil

An orifice located in the center of the iris, through which light penetrates to the eyeball. It is surrounded by a sphincter which causes its expansion (mydriasis) or contraction (miosis).

palpebral sulcus

The superior and inferior palpebral sulci are semicircular folds formed by the skin covering the eyelid.

eyelids

Two cutaneous folds, one superior and the other inferior, that cover the anterior part of the eyeball. The orbicular muscles of the eyelids, located inside the eyelids, and the levator muscle of the eyelids located in the superior eyelid, ensure their mobility.

lacrimal canals

Thin ducts which originate in the lacrimal points, turn inwards and terminate in the lacrimal sac. Their function is to carry tears and any foreign bodies from the conjuntiva through the nasolacrimal groove and deliver them to the inferior meatus of the nasal cavity.

lacrimal glands

Grape-like clusters of glands located just inside the orbit, superior and lateral to the eyeball. They contain tiny secretory ducts which secrete the key ingredients and most of the volume of the tears which bathe the conjunctival surfaces and all the external surface of the eyeball.

lacrimal sac

A cylindrical cavity that receives the lacrimal canal and is continued downwards by the nasolacrimal duct.

lacrimal caruncle

Reddish or pink eminence located in the internal palpebral commissure.

eyelashes

Small hairs which line the free edges of the eyelids. They defend the eye against foreign bodies. Among them are the orifices: the tiny sebaceous and sweat glands.

palpebral conjunctive tissue

The membrane or mucosa that covers the internal surface of the eyelids and extends backwards to cover the sclera.

iris

Ring-like, pigmented tissue whose muscles control the amount of light entering the eye. The hole in the center of the iris is the pupil. The iris is composed of smooth muscle innervated by, autonomic nerves.

sclera

A layer of connective tissue that surrounds the eyeball, except anteriorly, where it is surrounded by the cornea. The sclera has a whitish color and is not transparent.

lacrimal points

Two small orifices located above the lacrimal papillae, which are eminences located in the internal palpebral commissure.

nasolacrimal duct

A descending continuation of the lacrimal sac. It terminates in the nasal fossas, where the lacrimal secretions are delivered through the inferior meatus located under the inferior nasal concha.

FUNDUS OCULI

A representation of the surface of the retina as seen through the pupil, using an ophthalmoscope.

retina

The retina is the internal layer of the eyeball and lines two thirds of it. It is formed of nervous tissue similar to that of the brain and contains cells that receive light and convert it into nervous impulses which are transmitted to the brain.

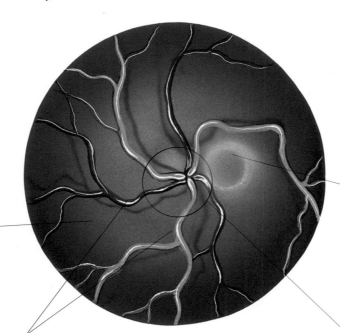

macula

Also known as the fovea centralis. A specialized area in the posterior of retina, rich in photoreceptor cells, where the image is focused.

arteries and veins of the retina

The retinal artery and vein reach the interior of the eyeball through the papilla parallel to the optic nerve. Inside the eye, they branch out over the entire retinal surface.

papilla

A circular, yellowish-white area, located in the posterior region of the retina, where it joins the optical nerve and where the central arteries and veins of the retina terminate.

150

VISION. OCULAR MUSCULATURE

medial rectus muscle of the eye

A muscle that crosses the internal face of the eyeball, from the common tendon of origin of the rectus muscles of the eye in the vertex of the ocular cavity and is inserted in the ocular sclera, a few millimeters inside the internal border of the cornea. It turn the cornea outwards. The action of this muscle in one eye is synchronous with that of the lateral rectus muscle of the other eye.

superior rectus muscle of the eye

A thin muscle located below and parallel to the levator muscle of the superior eyelid. It originates posteriorly in a tendon that serves as the common origin of the four rectus muscles of the eye and is attached to the vertex of the orbit. The muscle extends over the eyeball and is inserted in the sclera, a few millimeters above the superior border of the cornea. When contracted, it turns the cornea upwards and inwards, allowing the eye to move in these directions. Its contraction in one eye is coordinated with the inferior oblique muscle of the other eye.

superior oblique muscle of the eye

A muscle located in the roof of the orbit internally to the levator muscle of the superior eyelid. It originates posteriorly to the levator muscle and goes to the internal border of the orbital foramen where it is converted into a tendon which is attached to a ligament that makes an acute angle, passing below the superior rectus muscle of the eye. It is inserted in the superior internal area of the sclera and acts to turn the cornea downwards and outwards.

optical nerve

A thick nerve that enters the ocular cavity and penetrates the eyeball through its posterior face. The optical nerve sends visual sensations received by the retina and sent to the brain.

levator muscle of the superior eyelid

A flat, triangular muscle that crosses the vault of the orbital cavity from anterior to posterior. It originates in the bone, forming the functus of the orbit and is inserted in the subcutaneous tissue of the upper eyelid. It acts to elevate the superior eyelid.

151

orbicular muscle of the eyelids

A circular facial muscle surrounding the orifice, known as the palpebral orifice. It originates in the internal angle of the eye and is inserted in the external angle. It is attached to the skin of the eyelids in all its trajectory. It serves to open or close the eyelid.

inferior rectus muscle of the eye

The inferior rectus muscle has a common posterior origin with the other rectus muscles of the eye. It crosses the floor of the ocular cavity and is inserted in the anterior inferior part of the sclera of the eyeball, a few millimeters below the inferior border of the cornea. Its contraction turns the cornea downwards and inwards, directing the vision in this direction. Its action in one eye is coordinated with the superior oblique muscle of the other eye.

inferior oblique muscle of the eye

A muscle that crosses the floor of the ocular cavity from its internal or nasal part towards the external part. It originates in the bone, forming the floor of the orbit, crosses the inferior hemisphere of the eyeball below the inferior rectus muscle and is inserted into the inferior external area of the sclera. It turns the cornea upwards and outwards.

lateral rectus muscle of the eye

A muscle that crosses the external face of the eyeball, from the tendon that is the common origin of the rectus muscles of the eye in the vertex of the ocular cavity, to the ocular sclera, a few millimeters outside the external edge of the cornea. It serves to turn the cornea outwards when contracted. Its action is synchronous with that of the medial rectus muscle of the opposite eye ensuring both eyes turn in the same direction.

VISION. THE EYEBALL

cornea
A layer of connective tissue that covers the anterior part of the eyeball, causing it to protrude. The cornea is totally transparent, in order to let light pass.

iris
A structure which forms part of the intermediate layer of the wall of the eyeball. It is disc-shaped and located in the anterior face of the eye. The hole in the center of the iris is the pupil. It varies in color according to its transparency and vascularization, determining the color of the eyes.

ciliary body
An internal protuberance located between the iris and the choroidea, that contains the ciliary muscle and the ciliary processes, which are formed by a dense vascular network and secrete the vitreous humor. Its cross-section is triangular and its shape is annular, surrounding the iris externally.

suspensory ligament of the lens
Transparent fibers which unite the internal edge of the ciliary body with the periphery of the crystalline lens, maintaining it in its fixed position.

vitreous body
A viscous, transparent fluid that fills the ocular cavity located behind the crystalline lens.

sclera
A layer of connective tissue that surrounds the eyeball, except anteriorly, where it is surrounded by the cornea. The sclera has a whitish color and is not transparent.

choroidea
The intermediate layer of the three which form the wall of the eyeball, the choroidea is formed by a complex network of blood vessels that nourish the retina. It surrounds two-thirds of the eyeball and the rest is covered by the iris.

retina
The retina is the internal layer of the eyeball and lines two thirds of it. It is formed of nervous tissue similar to that of the brain and contains cells that receive light and convert it into nervous impulses subsequently transmitted to the brain.

papilla
A circular, yellowish-white area, located in the posterior region of the retina, where it joins the optical nerve and where the central arteries and veins of the retina terminate.

152

anterior chamber
The space between the cornea and the iris, which is occupied by vitreous humor.

pupil
An orifice located in the center of the iris, through which light penetrates to the eyeball. It is surrounded by a sphincter which causes its expansion (mydriasis) or contraction (miosis).

posterior chamber
A space located behind the iris and in front of the crystalline lens. It contains vitreous humor.

crystalline lens
An epithelial structure located behind the iris. Its anterior face is bathed by vitreous humor and the posterior face by the vitreous body. It functions as a biconvex lens whose changes of shape allow light impulses to be focused on the retina.

central artery and vein of the retina
Two blood vessels that reach the retina following the axis of the optic nerve and terminate in the interior of the eyeball through the papilla, forming branches which cover the retinal surface.

optical nerve
A thick nerve that unites the eyeball with the central nervous system, transporting the light sensations, converted in the retina into nervous stimuli, to the brain, specifically to the area of the occipital cerebral cortex in charge of consciously perceiving these sensations.

VISION. CONSTITUTION OF THE RETINA

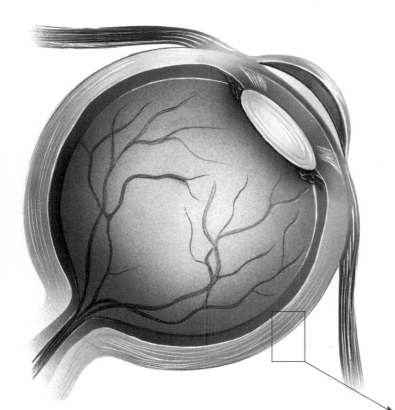

RETINA

The retina is the internal layer of the eyeball and lines two thirds of it. It is formed of nervous tissue similar to that of the brain and contains cells that receive light and convert it into nervous impulses which are transmitted to the brain. In the posterior zone of the retina, there is an area called the maculae containing many photoreceptor cells, which give clarity to the vision.

pigmentary epithelium
A layer of cells that produce a pigment called melanin, which protects and isolates the photoreceptor cells.

amacrine cells
Cells that associate and connect the bipolar cells and the ganglion cells, transmitting information about the light impulses received from one point of the retina to another.

photoreceptor cells
The photoreceptor cells are light-sensitive retinal cells (cones and rods), located below the pigmented epithelium. They contain chemicals called photopsin and rhodopsin which react to specific light wavelengths and trigger nerve impulses.

horizontal cells
Cells that fix the connections between the photoreceptors and the bipolar cells, transmitting information from the light impulses received in each point of the retina. Together with the amacrine cells, they are called association cells.

bipolar cells
Cells connected by a synaptic union with the photoreceptors. They capture the nervous impulses generated by the photoreceptor cells and transport them to the ganglion cells.

optical nerve
The optical nerve is formed by the axons of the ganglion cells and extends to the optical chiasm. After the two optical nerves join, it goes to the optical thalamus and posteriorly to the occipital cerebral cortex or visual cortex.

ganglion cells
Nervous cells that receive the impulses transmitted by the bipolar cells and carry them to the brain through a long axon which becomes part of the optic nerve.

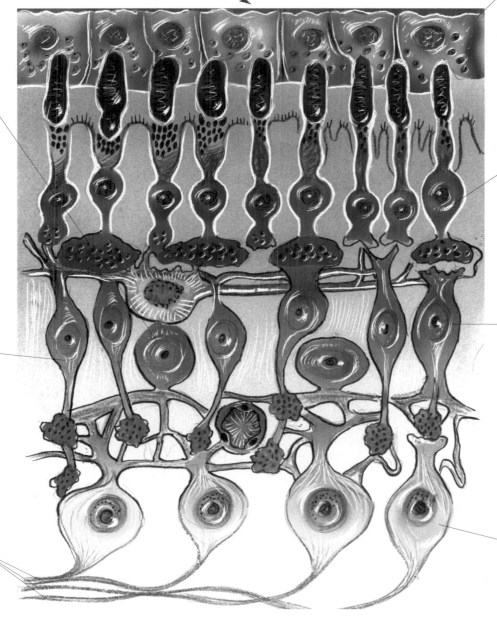

HEARING

pinna or **auricular pavilion**
The external, visible part of the ear, which surrounds the outer ear canal. It is principally composed of four cartilaginous structures: the helix, antihelix, tragus and antitragus.

helix
The curled rim of the outer part of the ear.

antihelix
Part of the cartilage of the ear which forms a curved elevation within or in front of the helix.

tympanic part of the temporal one
One of the parts of the temporal bone, whose horizontal area constitutes the roof of the external acoustic duct.

tympanic membrane or **eardrum**
A fibrous, elastic membrane that separates the external acoustic duct from the middle ear. Through its vibrations, sounds from the exterior are transmitted to the articulated ossicles of the tympanic cavity.

articulated ossicles
Three articulated bones (malleus, incus and stapes) which transmit the vibrations of the tympanic membrane to the labyrinth.

semicircular ducts
Three tubes arranged in three different planes which have receptors that capture the movements of the endolymph. They are essential for the maintenance of posture and balance.

internal acoustic meatus
A bony duct located in the petrous bone. It gives passage to the vestibulocochlear and facial nerves and the intermediate nerve of Wrisberg, connecting the cochlea with the interior of the cranial cavity.

cochlea
A spiral duct located below the vestibule. It contains the organ of Corti, which transforms auditory sensations into nervous stimuli.

vestibule
An elongated cavity containing articulated ossicles that communicates with the tympanic cavity through an orifice called the oval window.

labyrinth
A set of osseous cavities located in the sinus of the petrous bone, which correspond to the internal ear and contain membranous structures filled with a fluid called endolymph. There are three cavities which form the labyrinth: the cochlea, vestibule and semicircular canals.

peristaphyline muscle
A muscle inserted in the petrous bone and the proximities of the Eustachian tube and extending to the soft palate, which it tenses when contracted.

Eustachian tube
A duct that communicates the tympanic cavity with the pharynx and permits air from the nasal fossas to reach the cavities of the ear, thereby balancing the pressure on both sides of the eardrum.

internal jugular vein
The internal jugular vein originates in the union of the cranial venous sinus in the area of the petrous bone and receives the venous blood coming from the intracranial structures.

tympanic cavity
A cavity contained in the petrous segment of the temporal bone which houses the articulated ossicles. It corresponds to the part of the auditory system called the middle ear.

facial nerve
A sensory-motor nerve also known as the seventh cranial nerve which, after penetrating the internal acoustic duct, crosses the petrous bone by the Fallopian aqueduct and leaves the cranium through the stylomastoid foramen, dividing into temporal and cervical terminal branches.

external acoustic duct
Duct that communicates the pinna with the eardrum. It is covered by a prolongation of the skin of the pinna which, in turn, is covered by very fine hairs.

154

OLFACTION

nasal vestibule
A widening that constitutes the initial part of both nasal fossas. The vestibule, like the rest of the nasal fossas, is covered by a mucosa rich in mucous glands and small hairs which filter the air.

concha
A bony protuberance covered by nasal mucosa located in the lateral walls of the nasal fossas. It creates turbulence in the inspired air, thus warming and humidifying it before it reaches the pharynx

olfactory bulb

olfactory nerve
Also called the first cranial pair. The olfactory nerve is formed by the axons of the nervous cells that cross the cribiform lamina of the ethmoid bone from the olfactory mucosa in the nasal fossas and terminate in the olfactory bulb.

olfactory tract
A cord of nervous tissue that transmits the olfactory sensations captured by the nasal fossas to cerebral centers, which interpret them.

sphenoidal sinus
A cavity located in the sphenoid bone, which communicates with the nasal fossas through orifices located behind the conchae It serves to warm the air before it reaches the inferior respiratory tract.

155

olfactory glomerulus
The point where the olfactory nerves are united with the mitral cells, whose axons extend along the olfactory bulb and constitute the nerve fibers of the olfactory tract.

nerve fibers of the olfactory tract
The nervous cells of the olfactory bulb, called mitral cells, are united, on the one hand, with the olfactory nerves and, on the other hand, emit prolongations or axons that carry the olfactory sensations through the olfactory tract to the brain.

olfactory bulb
A swelling located at the end of the olfactory tract, over the cribiform lamina of the ethmoid bone. The nervous fibers that make up the olfactory nerves reach the olfactory bulb and are transmitted to the brain in the area of the hippocampus.

cribiform lamina
A part of the ethmoid bone located between the cranial cavity and the nasal fossas. It has a series of small orifices which allow the passage of branches of the olfactory nerve.

olfactory mucosa
Mucosa that covers the posterior superior part of the nasal fossas and contains the nervous cells specialized in capturing smells.

olfactory cells
Multiple glands dispersed between the olfactory cells of the nasal mucosa produce a mucous secretion.

olfactory gland
Nerve cells specialized in capturing smells. The olfactory cells are bipolar cells which emit a series of small cilia at one extreme which go to the nasal cavity, and at the other extreme, they continue, as nervous prolongations or axons, to form the olfactory nerve.

GUSTATION. TONGUE

LINGUAL PAPILLAE

Small papillae that cover the surface of the buccal mucosa and contain the taste buds. They can be of different types, according to their shape and function.

 area of bitter taste sensation

area of sour taste sensation

area of salt taste sensation

area of sweet taste sensation

terminal sulcus
A V-shaped sulcus that crosses the tongue transversally and delimits the superior or dorsal face of the tongue, descending vertically to form the lingual tonsils.

fungiform papillae
The fungiform papillae are distributed along the edges of the tongue and are bigger than the filiform papillae. They contain taste buds located in the superior face of the lingual papillae.

filiform papillae
The filiform papillae are simple elevations of the epithelium of the buccal mucosa, which constitute the smallest type of taste buds. They are very numerous and are distributed over the two anterior thirds of the tongue's surface.

circumvallate papillae
The large, prominent bumps on the top surface of the back of the tongue are a fourth type of papilla called circumvallate papillae. They are located along the circumvallate line and contain taste buds that allow the back of the tongue to distinguish sour and bitter tastes.

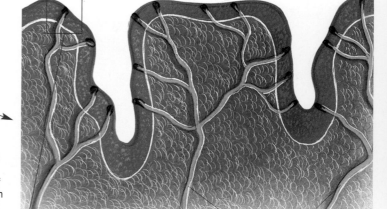

nerve fibers
Fibers emitted from each taste bud which, by means of different nerves which go to the brain, transmit the nervous stimuli generated by the taste receptors specifically to an area of the cerebral cortex called the operculoinsular region.

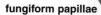

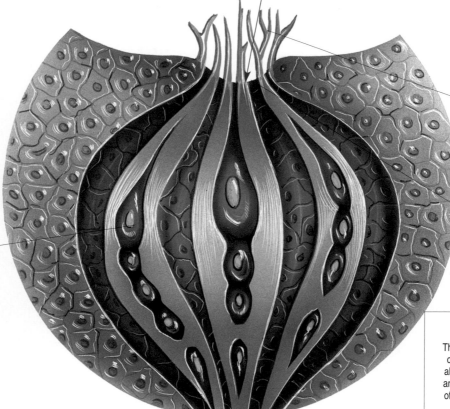

gustatory cells
Each taste bud contains around twenty gustatory cells whose function is to capture taste. They are stimulated by the substances responsible for a specific taste, which are dissolved in the saliva, and turn this sensation into a nervous stimulus.

microvilli
Tiny villi or taste hairs that line the external extreme of the gustatory cells. They show through the gustatory pore and provide the receptor surface necessary to capture flavors.

gustatory pore
An orifice in the epithelium that lines the buccal mucosa and communicates it with the cells located in the interior of the taste bud.

TASTE BUD

The taste buds, containing the gustatory cells, are mainly located in the tongue although they also cover the soft palate and the pharynx. They capture the taste of any substance ingested and transmit the information to the brain through the nervous fibers.

TOUCH. TACTILE CORPUSCLES

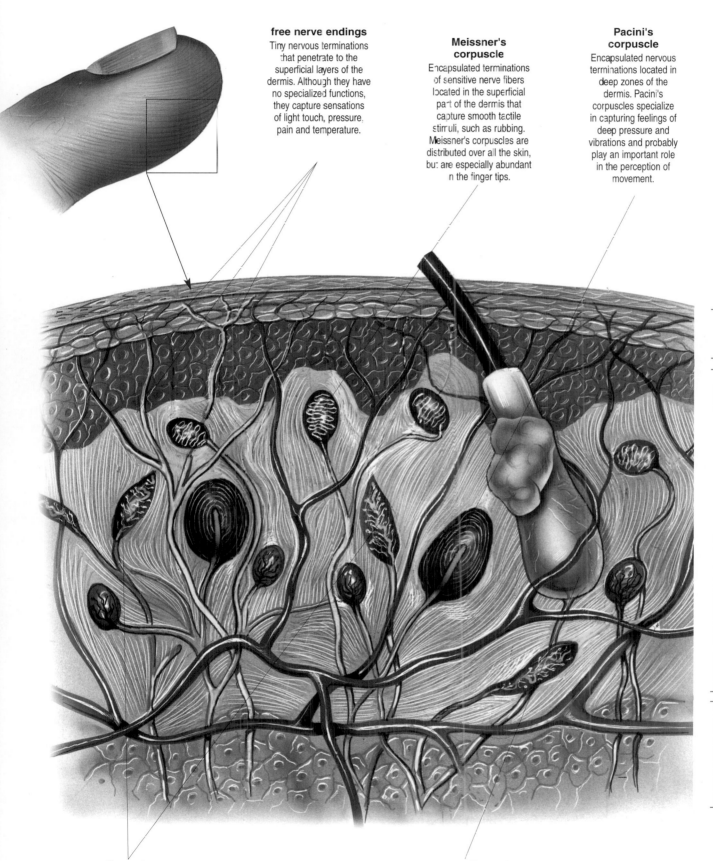

free nerve endings
Tiny nervous terminations that penetrate to the superficial layers of the dermis. Although they have no specialized functions, they capture sensations of light touch, pressure, pain and temperature.

Meissner's corpuscle
Encapsulated terminations of sensitive nerve fibers located in the superficial part of the dermis that capture smooth tactile stimuli, such as rubbing. Meissner's corpuscles are distributed over all the skin, but are especially abundant in the finger tips.

Pacini's corpuscle
Encapsulated nervous terminations located in deep zones of the dermis. Pacini's corpuscles specialize in capturing feelings of deep pressure and vibrations and probably play an important role in the perception of movement.

epidermis
The superficial, external layer of the skin, formed by layers of cells that are constantly renovated, with the dead keratinized cells flaking off to be replaced by others.

157

dermis
The intermediate of the three skin layers. Composed of loose connective and fibrous tissue, it contains many nerve terminations responsible for capturing tactile sensations.

hypodermis
The deepest layer of the skin, located below the dermis. It is formed of loose connective tissue and contains abundant adipose tissue, which acts as a cushion for the organs below (muscles, bones, viscera, etc.).

thermal receptors
Sensitive terminations in the dermis specialize in capturing thermal sensations such as cold and heat.

nerve endings of root hair plexus
Nervous terminations located in the hair root which capture any sensation experienced by the external part of the hair.